The Psychometrics of Standard Setting

This book provides a unifying structure for the activities that fall under the process typically called "standard setting" on tests of proficiency. Standard setting refers to the methodology used to identify performance standards on tests of proficiency. The results from standard setting studies are critical for supporting the use of many types of tests. The process is frequently applied to educational, psychological, licensure/certification, and other types of tests and examination systems. The literature on procedures for standard setting is extensive, but the methodology for standard setting has evolved in a haphazard way over many decades without a unifying theory to support the evaluation of the methods and the validation of inferences made from the standards. This book provides a framework for going beyond specific standard setting methods to gain an understanding of the goals for the methods and how to evaluate whether the goals have been achieved.

The unifying structure provided in this book considers policy that calls for the existence of performance standards, the relationship of proficiency test design to the policy, and tasks assigned to subject matter experts to help them convert the policy to estimates of locations on the reporting score scale for the test. The book provides guidance on how to connect the psychometric aspects of the standard setting process to the intentions of policy makers as expressed in policy statements. Furthermore, the structure is used to support validity arguments for inferences made when using standards. Examples are provided to show how the unifying structure can be used to evaluate and improve standard setting methodology.

Statistics in the Social and Behavioral Sciences Series

Series Editors
Jeff Gill, Steven Heeringa, Wim J. van der Linden, Tom Snijders

Recently Published Titles

Big Data and Social Science: Data Science Methods and Tools for Research and Practice, Second Edition
Ian Foster, Rayid Ghani, Ron S. Jarmin, Frauke Kreuter and Julia Lane

Understanding Elections through Statistics: Polling, Prediction, and Testing
Ole J. Forsberg

Analyzing Spatial Models of Choice and Judgment, Second Edition
David A. Armstrong II, Ryan Bakker, Royce Carroll, Christopher Hare, Keith T. Poole and Howard Rosenthal

Introduction to R for Social Scientists: A Tidy Programming Approach
Ryan Kennedy and Philip Waggoner

Linear Regression Models: Applications in R
John P. Hoffman

Mixed-Mode Surveys: Design and Analysis
Jan van den Brakel, Bart Buelens, Madelon Cremers, Annemieke Luiten, Vivian Meertens, Barry Schouten and Rachel Vis-Visschers

Applied Regularization Methods for the Social Sciences
Holmes Finch

An Introduction to the Rasch Model with Examples in R
Rudolf Debelak, Carolin Stobl and Matthew D. Zeigenfuse

Regression Analysis in R: A Comprehensive View for the Social Sciences
Jocelyn H. Bolin

Intensive Longitudinal Analysis of Human Processes
Kathleen M. Gates, Sy-Min Chow, and Peter C. M. Molenaar

Applied Regression Modeling: Bayesian and Frequentist Analysis of Categorical and Limited Response Variables with R and Stan
Jun Xu

The Psychometrics of Standard Setting: Connecting Policy and Test Scores
Mark D. Reckase

For more information about this series, please visit: https://www.routledge.com/Chapman--HallCRC-Statistics-in-the-Social-and-Behavioral-Sciences/book-series/CHSTSOBESCI

The Psychometrics of Standard Setting

Connecting Policy and Test Scores

Mark D. Reckase

CRC Press is an imprint of the
Taylor & Francis Group, an **informa** business
A CHAPMAN & HALL BOOK

First edition published 2023
by CRC Press
6000 Broken Sound Parkway NW, Suite 300, Boca Raton, FL 33487-2742

and by CRC Press
4 Park Square, Milton Park, Abingdon, Oxon, OX14 4RN
CRC Press is an imprint of Taylor & Francis Group, LLC

ISBN: 9781498722117 (hbk)
ISBN: 9781032422619 (pbk)
ISBN: 9780429156410 (ebk)

DOI: 10.1201/9780429156410

Typeset in Palatino
by Deanta Global Publishing Services, Chennai, India

This book is dedicated to Charlene Repetny Reckase who set the standard for me for what it meant to be a good and generous human being.

Contents

Preface

I formally started working on this book in September 2014. Although it seems that eight years is a long time to work on a book, and it is, the development of the ideas presented in this book began much earlier, in 1992. Prior to that year, I not only knew nothing about the process of setting standards on educational and psychological tests, but I also carefully avoided learning anything about that topic. My perception of myself as a psychometrician led me to naively dismiss work on standard setting as too subjective to be worthy of serious consideration. Of course, my naive judgment was wrong, but I needed an important event to get me to see the inaccuracy in my thinking.

In 1992, I was Assistant Vice President for the Assessment Innovations Area at ACT, Inc. In that year, the National Assessment Governing Board (NAGB) awarded ACT, Inc. a contract to develop performance standards on the reporting score scales for the National Assessment of Educational Progress (NAEP) and my direct supervisor, the Vice President of the Assessment Division at ACT, Inc. Cyndie Schmeiser, asked me to work on the contract. As I recall the event, admittedly a fuzzy memory after all these years, I tried to avoid that assignment by indicating that I knew nothing about standard setting. My memory is that she said something like, "No one else knows anything about standard setting either. We want you to work on this." I understood that I was not really being "asked" to work on the project, I was being assigned to work on the project. The fact that this book now exists should be credited to Cyndie Schmeiser.

I worked on that contract, and subsequent renewals, for about ten years. Over that period of time, I learned about standard setting, and the ideas presented in this book were developed through conversations with project staff, especially Susan Cooper Loomis; the project director, Luz Bay, I think she was the assistant director; and Ric Luecht and Tom Hirsch, who were psychometricians for the project. There was also a technical advisory committee that included Ron Hambleton, Bill Mehrens, and Wim van der Linden. Their discussions of the details of standard setting procedures were extremely helpful in making me understand the nuances of standard setting processes. It is important to indicate that I have not asked any of them to endorse what I have written in this book. They may or may not agree with the way that I think about the process of setting standards on tests.

I also need to give credit to John Kimmel, who was the original editor for this book. He was unwaveringly patient as I worked on the book and paused to care for my wife, who developed Alzheimer's disease in the middle of the writing process. She was one of my best editors. She read through my entire multidimensional item response theory manuscript before that book was published. She provided many suggestions for improvements. I believe that this book would be much better if she were able to give her valuable input. It is unfortunate for me that John retired before I could finish the manuscript. I hope that retired life is working out well for him. Lara Spieker is now the editor for the book. I appreciate her continuing patience as I work to complete this project.

There are many other people I should recognize as providing reviews and good advice about the book. I fear that I cannot remember all of them given the length of time I have been working on the manuscript. My son, Erik Reckase, who works in an entirely different field, is one of them. I am amazed that he was willing to read these long chapters in draft form. His input is greatly appreciated. And he did find weaknesses in logic and inconsistencies in notation along with some editorial problems. It is a great benefit to have a reviewer who is not so immersed in the field that they read into the text things that are not there. There are many others. Jing-Ru Xu and Yanlin Jiang read several of the chapters and provided insightful comments and helpful editorial suggestions. I have also had thought-provoking conversations with Dan Lewis, who helped me refine the connection between policy and standards on score scales. There were also comments by anonymous reviewers who John Kimmel recruited that were very helpful. I think I know who they are, maybe Adam Wyse and Brian Clauser, but I am not certain. I hope they are not offended by my including them here.

There are many others who unknowingly contributed to this book. They are colleagues and contractors, state and federal testing staff, and persons at professional meetings who described their views of the standard setting process and some of the controversies around them. There have been many stimulating conversations over the years that helped me refine the ideas presented in this book. I am sure that some of those individuals will not agree with the framework that I have presented in the book. I hope they will find those disagreements as motivation to present their own view of the theory underlying standard setting. I will look forward to reading those works.

Mark D. Reckase
Okemos, Michigan
July 11, 2022

Author biography

The author of this book has decades of experience developing and reviewing standard setting processes as well as doing advanced work in psychometric theory. He is currently a University Distinguished Professor Emeritus at Michigan State University. He has been the editor of *Applied Psychological Measurement* and the *Journal of Educational Measurement*. He has also served as the president of the National Council on Measurement in Education and the International Association for Computerized Adaptive Testing, and as the vice president of Division D of the American Educational Research Association.

1

Conceptual Overview for a Theory of Standard Setting

1.1 The Prevalence of Standards and Standard Setting

Standards are so prevalent in our everyday lives that we hardly notice them. When I had breakfast this morning, the side of the cereal box indicated that I was getting 25% of the daily requirement for vitamin B12. That information implies that someone somewhere had determined the daily amount of that vitamin that was required to maintain good health for the adult population in the United States. Few people who read the side of the cereal box know (or even care) how this nutritional standard was derived, and I believe most people accept the information as accurate. When I drove to my office this morning, the speed limit was shown on a sign at the roadside as 45 miles per hour. Someone someplace had determined that that was the appropriate maximum speed of travel on that road, but most of us do not know who was involved in setting that limit or the process that yielded that result. Generally, we accept the speed limit as reasonable and appropriate without knowing all the details about how the standard was developed. It would be easy to identify many other examples of standards that we encounter in our daily lives. Most of these standards go unnoticed and they are accepted without critical evaluation. There is an assumption that there is sound logic behind the standard and a good process was used to determine the specific value that is recommended.

Some standards have a more dramatic impact on our lives than the information on the side of a cereal box, so they are given closer scrutiny. Many occupations require that candidates for jobs show that they have an acceptable level of competency before they are hired. For example, employment as a registered nurse in the United States usually requires that candidates show that they have acquired knowledge and skills that are deemed to be important for successfully performing job requirements in the nursing profession (Dorsey & Schowalter, 2008). More broadly, Cusimano (1996) argues that criterion-referenced standards, those requiring evidence that a candidate has acquired specific knowledge and skills,

DOI: 10.1201/9780429156410-1

are needed to protect the public when hiring individuals for the medical profession. The usual way for job candidates to demonstrate that they have the required knowledge and skills is for them to take a test composed of tasks that require the successful use of a sample of the important skills and knowledge. If the candidate achieves at or above a specified minimum level of performance on that test, he or she is inferred to have sufficient knowledge and skills to work in the profession. The specified minimum score on the test is the operationalization of the standard for the required amount of knowledge and skills.

Because the standard used with a test of competency for an occupation has direct consequences for each individual taking the test, many of the individuals who take such tests are rightfully concerned about whether the standard is set at an appropriate level. They want to know how the standard was set and what evidence exists that supports the appropriateness of the specified level of performance. The level of interest in the process for setting standards related to employment is shown by the amount of case law that relates to employment testing and the appropriateness of standards (see Phillips (2010) for numerous examples).

Because a policy was implemented in 1990 requiring the setting of performance standards on the National Assessment of Educational Progress (NAEP) (see Bourque (2009) for details of the implementation) and because of subsequent federal legislation requiring that states have achievement tests in certain subject matter areas and standards of proficiency on those tests (see Linn, Baker and Betebenner (2002) for a summary of requirements), interest in performance standards set on tests and the process used to determine them has become more widespread. Teachers, schools, and individual students have been evaluated using proficiency standards on state-mandated standardized achievement tests to determine if the educational system is performing as desired (Hunter (2011), Michigan Department of Education (2010), National Center for Education Statistics (2014), etc.). Because of these uses of tests and standards, the broader community of educators and those actively involved in education have become more interested in proficiency standards on tests and how they are set. As a result, there has been a substantial amount written about the process of setting standards of performance on tests of proficiency and the ways that these standards can be evaluated to determine if they are credible (see Cizek (2012), Hambleton & Pitoniak (2006) for example).

Despite the large number of journal articles and books about the setting of performance standards on tests, there has been little effort to specify a formal theory to support the standard setting process. Instead, many ad hoc procedures for standard setting have been developed. Glass (1978) was very critical of how these procedures had been developed. "To my knowledge, every attempt to derive a criterion score is either blatantly arbitrary or derives from a set of arbitrary premises" (p. 258). However, since his 1978 article, these "arbitrary" procedures have been used and studied to determine how they can best be implemented. In most cases, the lack of a formal theory of standard setting has not caused serious problems because there is substantial practical experience that shows that various methods give reasonable results and there have been legal cases that have tested the support for the reasonableness of results. In a sense, the practice of standard setting is much like the practice of constructing buildings before the science of strength of materials was developed. Master carpenters learned through an apprenticeship how to build sound structures even though they did not understand the chemical and physical properties of the materials that they were using. However, as structures became larger and more complex, experience and rules of thumb were no longer sufficient to ensure that the structures were safe and would remain standing for the desired length of time. I would not want to drive my car across a bridge that was designed and built using only the best intuitions of the construction workers. Similarly, the detailed procedures for implementing standard setting processes are typically learned through activities like an apprenticeship. The approach taken for standard setting in a particular situation is typically based on the experience of the persons asked to perform the task. Little theory is called upon to guide the work.

The purpose of this book is to address what is seen as the arbitrariness of standard setting procedures by presenting initial work on a theoretical framework for the process of setting performance standards. Many of the theoretical foundations presented here apply to all types of standards from those related to nutrition to the speed limits on streets. However, because this work was stimulated by the attention placed on standards for entry into professions and those used to evaluate performance of educational systems, the theoretical foundations will focus on the connection between measures of performance, usually tests, and the standards set on the reporting score scales for those measures. In many ways, this is a more complex problem than setting standards for the amount of dietary components, or the recommended speed on highways, because those examples have existing measurement scales on which to set the standard. Educational, psychological, and workplace standards are typically specified on the scale defined by a test and the test design may have a direct influence on the process for setting the standard. Therefore, along with a basic theory of standard setting, there is also a psychometric theory of standard setting that addresses the needed characteristics of the test scores that are used as the measures of performance.

The remainder of this chapter gives a general overview of the topics that will be covered in this book. The purpose of the overview is to give intuitive descriptions of features of the standard setting process that will be given formal definition in later chapters. The following chapters provide formal definitions and a theoretical framework that indicates how the component parts of a standard setting process are interconnected and how a standard setting procedure can be evaluated.

1.2 Definitions of a Standard

The dictionary definitions for the word "standard" give some insights into the cultural context for setting standards. *The Random House Dictionary of the English Language (2nd edition) Unabridged* (1987) lists 28 definitions for the word "standard." A number of these definitions that are relevant to the educational and psychological context are the focus of this book.

> 1. something considered by an authority or by general consent as a basis for comparison; an approved model. 2. an object that is regarded as the usual or most common size or form of its kind. 3. a rule or principle that is used as a basis for judgment. 7. the authorized exemplar of a unit of weight or measure.

These definitions are useful, but more elaborated and nuanced definitions are needed for the purposes of this book. For example, definition 1 begins with "something considered by an authority." This phrase indicates that in many cases standards do not spring into existence on their own; they are called for by "an authority." For the applications considered here, the authority might be a certification or a licensure agency, or it might be the state board of education for a state, or a governmental agency such as the National Assessment Governing Board. In this book, the organization that calls for a standard will generically be labeled as the "agency." Agencies differ in their characteristics and the way they initiate and interact with standard setting processes. The role of the agency is an important consideration in a theory of standard setting.

The second part of definition 1 is that standards form the basis for comparisons, or an approved model that can be used for comparison. In most cases where educational or psychological tests are used, the basis for comparison is a score on a score scale, but it may be thought of more generally as a person with the desired characteristics. In standard setting studies, there is often reference to the "minimally qualified person" who is the "approved model" to which other individuals are compared. The same minimally qualified person can be thought of as the "authorized exemplar" in definition 7 and the "basis for judgment" in definition 3.

Of those listed, definition 2 is related to what is called "norm-referenced" in the testing industry. That is, the performance of an individual is compared to the distribution of performance for a meaningful group of individuals. The comparison may be to the mean, median, or mode of the distribution, or some other percentage point, such as the 90th percentile. This may seem like an objective way to define a standard, but comparison to a distribution of performance still requires that someone (the agency?) decide the boundaries of who are included under the label "its kind" and the point in the distribution used as the standard.

In certain cases, the development of the standard is not initiated by a formal agency, but it is defined by the general sense of the use of words by a population of people. For example, in common usage, a person might be labeled as "tall." The specific standard for "tall" is set by the impressions of people and generally would be called "norm-referenced" in the same sense as definition 2. However, the clothing industry may have a trade organization that can be considered the agency that calls for a formal standard for the concept "tall" that can be used to uniformly specify the way clothing sizes are labeled. Care must be taken to distinguish informal standards derived from common usage and formal standards that are called for by an agency. This book focuses on the development of formal standards.

Overall, the definitions suggest that a standard is something called for by an agency that will serve as the basis for comparison for making judgments of quality. The standard may include an exemplar as a basis for comparison and there is usually some type of measurement scale (see definition 7) that facilitates determining if something or someone is above or below the standard.

The dictionary definitions of a standard do not provide any details about how a particular standard is determined. A standard is somehow determined by authority (the agency) or general consent. In this book, a distinction is made between the conceptual standard considered by authority or general consent and the operationalization of the standard on a measurement scale. Kane (1994) has made a similar distinction. Tall might conceptually mean being able to reach a high place, but the operationalized value might be 6 feet 5 inches (1.96 meters) or more in height. The term "standard setting" generally refers to the process used for determining the operational value that is consistent with the conceptual standard defined by the agency. This may involve identifying exemplars and using a measurement scale so that comparisons can be made to the location of exemplars on the scale. For a standard to be credible, the process of standard setting needs to be done in such a way that there is clear evidence that the standard has the meaning that is intended by the agency that calls for its existence.

The dictionary definitions do not make a distinction that is common in the design and development of educational tests. Definition 2 is consistent with the term "norm-referenced" in the measurement literature. However, the alternative term from the measurement literature, "criterion-referenced," is only suggested by the "authorized exemplar" in definition 7. The term "criterion-referenced" suggests that the standard is based on requirements for specific skills and knowledge, and these requirements are described independently of the summaries of performance of the entire population of interest. It is possible that almost all members of a population might match the described skills and knowledge or hardly any of them. As will be discussed later, many standard setting processes are labeled as criterion-referenced.

The two different kinds of standards - criterion-referenced and norm-referenced - tend to be described in different ways. Criterion-referenced standards have specific descriptions of the capabilities of the persons who meet the requirements of the standard and there is no implicit consideration of the number of persons who will exceed the standard when the specific value of the standard is determined. A simple example of a criterion-referenced standard for the concept "tall" is that persons meeting the standard can reach objects on the top shelf of a kitchen cabinet without any aid such as a stool or tool for extending reach. "Reaching the top shelf" is the criterion in the definition and there is a direct test for determining if a person meets or exceeds the definition of "tall." The standard setting method is the procedure for determining how high the top shelf is and what height, on average, is required to reach that high. Note that there is no consideration about how many individuals will meet or exceed the operational standard for "tall."

The norm-referenced type of standard is based on some point in the distribution of the characteristic of a group, such as an average or a specified percentage point. The conceptual definition of the norm-referenced standard for "tall" might be that the height of a person is relatively rare in the population of persons who are of interest. The standard setting process focuses on operationalizing the meaning of "relatively rare" so that a height value can be determined that corresponds to it. That process might conclude that being in the top 10% of the population matches the meaning of "relatively rare." This approach would result in a standard for tall that is about 6 feet 1 inch (1.85 meters) for adults in the United States.

Standards are seldom strictly criterion-referenced or strictly norm-referenced. Often, in the process of setting standards, there will be a description of skills and knowledge, but also some information about the percent of individuals who meet or exceed that description. The agency calling for the standard and/or the developer of the standard setting process need to specify the amount of emphasis given to criterion-referenced versus norm-referenced information when implementing the process for determining the location of the standard on the measurement scale. A source of confusion is that a standard setting process may be labeled as criterion-referenced, but still provide information about the observed distribution of performance for the target population during the process of setting the standard. Different standard setting implementations that are given the same label may use different combinations of criterion-referenced and norm-referenced judgments. This leads to ambiguity in the meaning that can be attributed to the specific value of a standard and the types of supporting evidence needed to support the credibility of the standard. Is a person labeled "tall" because they exceed the 90th percentile for a population, or because they can perform a task that only persons of a specified height can do? These variations and ambiguities imply that a theory of standard setting is needed that can guide the way that standard setting processes are implemented, and results are reported and interpreted.

1.3 The Differing Contexts for Setting a Standard

As noted previously, the term "agency" will be used in an abstract way to refer to those who call for a standard. This usage will facilitate the development of formal theory. However, the reality is that the agencies that call for a standard vary in their composition and goals. They function within differing social/political contexts. These variations in structure of the agency and context for the standard setting have strong influences on the way that the process is carried out.

One of the ways that the agency or context differs is the breadth of the domain that is considered when calling for a standard. The example given earlier about the standard for the term "tall" is narrowly focused on a very specific measurement: height. When standard setting processes are applied to tests, the topic of the tests can be fairly narrow or very broad. In some cases, the agency calls for a standard on narrowly defined competency related to a specific kind of training. Emergency medical technicians (EMTs) are trained on very specific medical procedures, and the standard that they are required to meet is focused on the specifics of that training. A high school graduation requirement for competence in mathematics is much more broadly defined because the use that students will make of the acquired skills and knowledge in mathematics is unknown at the time a test is administered. In the first case, there is a fairly narrow body of knowledge and skills that is required, and the standard is used to make critical decisions about whether a person will be allowed to work in the profession. Shepard (1980) indicated that this type of standard is not for the benefit of the examinee but is used to protect some larger public. This is informally called a "high-stakes" use for the standard.

In the second case, the body of knowledge and skills is broad because the way students will use mathematical content is unknown. Some students may go on to careers that regularly use mathematics, while others may go on to ones where mathematics is less critical. One way of dealing with this issue is to consider the standard as determining whether the students have acquired enough mathematics to take the next instructional program in the sequence: entry-level college mathematics. The mathematics needed for those going directly into the workplace may not be considered because there are too many different types of employment and they require vastly different levels of skills and knowledge in mathematics.

Often educational performance standards are not used to make decisions about individual students but may have consequences for teachers or schools. In those cases, Shepard (1980) suggested that these standards are for program evaluation rather than individual evaluation. The stakes for students may be low while the stakes for teachers and schools may be high to moderate. Educational standards often have the added complexity of defining multiple levels of performance. Standards on the NAEP classify student performance into four different levels called achievement levels (e.g., National Center for Education Statistics, 2014).

This brief discussion has already suggested a number of ways that the context for the standards may vary: (1) whether stakes are for the individual taking a test, or for some person or aggregation of persons with a more remote connection to performance on the test, such as a teacher or a school; (2) whether the knowledge and skills required are focused on the specifics needed for an occupation or task, or if they are more generic such as the requirements for compulsory education; (3) whether the stakes are high, indicating some critical decisions are made using the results, or low, indicating that the results are more informational or advisory; and (4) whether one standard is needed or there are multiple levels of standards.

The specifics of the functioning of the agency and the context for the standard often determine the type of standard setting process that is used and the level of information that is needed to support the credibility of the standard. As stated by Shepard (1980), "It is not a new idea, then, to say that technical issues including standard setting will be influenced by the purpose of testing" (p. 449). Clearly, this issue has been with us for some time. It was not a new idea in 1980 and it is still with us.

High-stakes standards for those going into occupations may be contested because a person is not able to get a job and the standard may be considered unfair. In such cases, convincing support is needed for the decisions made using the standards. In other cases, there are no direct consequences to the use of the standard, so there is less pressure to obtain supporting information. Some agencies have substantial resources for developing a standard setting process, while others develop the standards with almost no resources. All these factors lead to the development of an amazing number of different standard setting procedures. Some of these are described in Chapter 5. More comprehensive listings of standard setting procedures are given in dedicated books on the topic, such as that edited by Cizek (2012). Standard setting procedures are often customized to match the resources available and the criticality of uses.

1.4 Who Defines the Standard?

In this text, it is emphasized that an agency calls for the existence of a standard when a formal standard setting process is used. The agency can also determine the specific role it wants to play in working out the details of the standard setting process. There are two general philosophical positions in the literature on standard setting about how the conceptual definition for a standard of performance is determined. One position is that the agency that calls for a standard identifies qualified individuals, and it is the considered opinion of those individuals that constitute the standard. Clauser, Clauser, and Hambleton (2014) state in the context of an item rating method "The individual judgments provide evidence for estimating the judge's expert opinion" (p. 22). And, "This value can be unique for each judge." For example, if it is desired to have a standard of competence for those who perform as oboists in a symphony orchestra, experts on that type of musical performance can be identified and whomever they determine is competent is accepted as being above the level defined by the selected group of experts as the standard of performance. The standard setting process involves selecting the individuals who are qualified to give judgments on the quality of performance, and then letting them indicate who they judge as meeting their own internal criteria. I once heard a government administrator say that a special process for setting standards on educational tests was not necessary because he would indicate which students were proficient and which ones were not. I facetiously call this the "great person method." All judgments of proficiency are made by this one person because he "knows" the truth about the capabilities of students.

The other philosophical position in the standard setting literature is that the agency that calls for the standard defines the standard by the way that the agency describes it. Cizek and Bunch (2007) state it this way: "standard setting can be defined as a procedure that enables participants using a specified method to bring to bear their judgments in such a way as to translate the policy positions of authorizing entities into locations on a score scale" (p. 19). The agency may call on experts to help with the crafting of the description, but ultimately the agency approves the description as the definition of the standard. The standard setting process is a method for translating the description of the standard to a more precise framework that allows the practical determination of the persons who meet or exceed the standard. For example, the agency calling for the standard may describe it as "the amount of protein needed in an adult's daily diet to avoid any related health problems." The standard setting process translates this description into a numerical value on a measurement scale for the amount of protein such that those who have less protein than the stated amount are expected to have health problems related to the deficiency. There should be experimental/observational support for the specified amount of protein specified as the standard rather than relying on the subjective opinion of experts about whether a person meets the standard.

1.5 Qualifications of the Persons Who Participate in the Standard Setting Process

All procedures for setting standards require judgment. The judgment might be to identify people who match the requirements for being minimally qualified on the target construct. It might be a judgment about how someone who is minimally qualified will perform on a task on a test. There are many other types of judgments that are involved, such as deciding whether to use a mean or median for computing the standard. In all cases, persons are making these judgments and an important consideration is the qualification needed to make credible judgments. These persons also need to understand the context for the standard and the specific process that will be used to collect their judgments.

In some cases, the persons involved do not have to meet the standard themselves to provide meaningful judgments. A somewhat silly example is that a person does not have to be tall to tell if another person can reach the top shelf of a cabinet. But they do have to be capable of making accurate observations and have a good vantage point for observing a person's ability to reach the shelf to be able to tell if a person meets the requirements for the definition of "tall."

Some standard setting processes require that persons make judgements about the expected performance on test items of minimally qualified examinees. It is difficult to imagine that the judgments will be credible if the persons making the judgments cannot understand the content of the questions themselves. Persons with different qualifications are needed depending on the details of the judgment task for different standard setting processes. Part of the design of a standard setting process is determining the qualifications needed for the participants in the process. It does not make sense to ask persons who do not know anything about statistics to make a judgment about whether a mean or median should be used to determine the point on the continuum underlying the standard. Similarly, setting a standard for competency for an automotive mechanic is likely to require that those who make the judgments know something about automotive mechanics. A critical question is how much should a person who is a participant in a standard setting process know about the content that defines the continuum for the standard? Raymond and Reid (2001) give some guidelines for selecting persons to participate in the standard setting process.

Before setting standards on state assessments became a major component of the interpretation of the test scores, the persons who were invited to participate in a standard setting process were typically unfamiliar with the standard setting process. That meant that part of the standard setting process was training the individuals in the tasks that they were asked to do. This means that another criterion for selecting individuals to be part of the standard setting process is that they can learn from the training and be able to implement the process in a way that was consistent with the training. It sometimes happens in a standard setting session that a participant does not understand the procedure and may give information that is unrelated to the task or that even adds misinformation to the process. Part of the challenge when running a standard setting process is to determine what to do with the input from persons who do not seem to understand the procedures they are supposed to follow. Cizek (1996) discusses the issue of when panelists should be removed from the process because the information they provide reduces the credibility of the standard.

With the increased number of standard setting studies, it is now often the case that the individuals involved have participated in previous studies, possibly using a different standard setting method. The fact that the persons are experienced with a standard setting process can sometimes expedite the current study. However, when the person is familiar with a different standard setting process, the previous experience can interfere with the proper use of the new procedure. Those who run standard setting studies need to plan for this possibility.

Ultimately, the ideal participant in a standard setting study is a person who knows the content of the test, who can quickly learn the standard setting tasks and implement them with fidelity, and who is flexible in addressing the issues related to estimating the standard. Part of the standard setting process is finding persons with the desired characteristics. The political aspects of standard setting may impose additional requirements such as gender balance, regional balance, and representation of important constituents. Meeting all these requirements takes planning and sufficient time and resources for recruiting. In some cases, it may be necessary to include individuals in the standard setting process that do not meet the ideal characteristics. It may be that the ideal participants do not exist, or other factors make it important to include individuals without important skills or knowledge. In those cases, it is important to determine how those individuals can meaningfully contribute to the process, and also to minimize any error that they might add to the estimation of the standard.

1.6 The Standard Setting Process as a Data Collection Design

Ultimately, a standard setting process is a way to collect information (data) that can be used to estimate the location of a standard on a continuum. One of the reasons that there are many kinds of standard setting procedures is that there are many different types of information that can be collected and many different ways that the information collection can be organized. For example, one way of collecting information to define a standard is to identify a group of individuals that embody the minimum characteristics of the standard. Then that group is assessed on the target construct to acquire the data that are used to estimate the standard. A summary statistic for that group is used to define the location of the standard on the continuum for the construct. The example given earlier for the definition of "tall" can be used as a concrete example. Suppose a group of individuals is identified who can just reach the top shelf in a cabinet. These are the individuals who meet the minimum requirement for the standard for "tall." Then the mean or median of the heights for those in this group can be computed and used as the indicator of the point on the height continuum that is the minimum criterion for being considered tall.

The data collection design in this case is to identify enough individuals who can just reach the top shelf of the cabinet and then measure their heights. The results are analyzed using a measure of central tendency to estimate the point on the height continuum that corresponds to the standard. The size of the sample of individuals that is identified is related to the desired level of accuracy for the estimation of the standard. The sample size required will depend on the amount of variation in height for the group of individuals that were identified through the judgment process used for selecting the individuals. There will likely be some variation in height because of variation in length of arm, a factor in reaching the top shelf that is not explicitly included in the definition of height. Height is typically measured from the floor to the top of the head. It is important that the conceptual definition of tall provided by the agency consider the importance of variation caused by differing arm lengths.

Data collection designs for many standard setting processes are much more complicated than the example for "tall." Each standard setting process asks individuals to give judgments of a different kind, but it must be possible to use the information (data) provided in the judgments to estimate a point on the score scale used to represent the continuum of interest. The critical feature when thinking about standard setting as a data collection design is that there must be some procedures for estimating a point on the continuum that corresponds to the judgments that are collected. It is also important that it be possible to judge the accuracy of the estimate of the point on the continuum.

A significant part of this book is related to the requirements of data collection designs and to the statistical procedures that are used to determine the point on the continuum that is considered the final product of the process.

1.7 Calculating the Value of the Standard

The ultimate result of a standard setting process is usually a number that refers to a location on the scale used for reporting results. In the educational context, that scale is usually the reporting score scale for a test of achievement for students enrolled in an educational system. In the licensure/certification context, the location is on the reporting score scale for the examination used to make the judgment about the level of competence for performing the tasks required in a profession. In our fanciful example of setting a standard for "tall," the location is on a scale for measuring height.

If we had perfect judgments and high-precision measurements, the determining of the location on the scale would be straightforward. If all the persons who could reach the top shelf of a cabinet were the same height, and all observers agreed that the same persons reached the shelf, then the computed value for the standard for height would be the same for all observers. In reality, however, all of the persons would not be the same height because of differences in the geometry of arm length and body size, and there may be disagreements among the observers about what is meant by being able to reach the shelf. Given that there is likely to be variation in the data collected to represent the standard, some method must be selected to arrive at a single number that represents the standard. The many complexities for this simple idea will be discussed in detail in a later chapter.

In this introductory discussion, it is helpful to consider whether the determination of the location on the reporting score scale is called estimation or simple computation. Estimation implies that there is at least a conceptual true location and the goal of determining the location for the standard is to try to get as close to the true location as possible. For example, suppose that rather than defining "tall" as being able to reach a high shelf, we define "tall" as being above some height measure that is specified as the most frequently mentioned minimum value for being considered tall by a population of adults in society. In theory, we could ask every person in the population for their minimum height measure for "tall" and then determine the mean of the distribution of the specified values and define the result as the true value for the definition of "tall." In practice, however, we can only ask a sample of persons, rather than everyone in the population, and compute a measure of central tendency as an estimate of the population value. If the computed value is close to the true value, it is a good estimate. The goal in this case is to accurately estimate the population value.

An alternative view is that there is no population parameter to be estimated so the value that is computed is also a matter of judgment (see Clauser, Clauser, and Hambleton, 2014 for example). Any value that is accepted by the agency that calls for the standard as matching their intentions is considered correct. The quality of the standard is based on the acceptance by the agency of the process, the consistency of judgments by panelists, and the computational method.

A third view is that there is a true value for the standard, but it is not defined in a statistical way as a parameter of a distribution. Instead, it is defined as the intention of the agency when they call for the standard. In a sense, the description of the meaning of the standard is in one language, and the value of the standard on the reporting scale is in another language. The goal of the computation of the standard is to accurately translate from the first language to the second. In this case, the computational process used to compute the location on the reporting score scale from the judgments that are collected needs to provide an accurate translation of the agency's intentions. Checking the translation would be somewhat like having two persons who are competent in both languages perform the translation and then determining if they obtain the same result. In a very abstract sense, it might be argued that there is a "true" or "best" translation.

1.8 Evaluation of the Results of a Standard Setting Process

At the end of a standard setting process, one or more locations are specified on a continuum that is the operationalization of the construct that is the target for the standard. In educational settings, the locations are specified by test scores on the score scale used for reporting the results on an achievement test that is the operationalization of the content construct, which is the target for the process. A reasonable question about the results of the process is whether the results are credible and/or have evidence for the validity of inferences drawn from the use of the standard. That is, when a person is estimated to be above the location on the reporting score scale, does that person really have the characteristics implied by exceeding the standard as described by the agency? That the person does have those characteristics is the basic inference implied by the use of the standard to make decisions. This inference would be the focus of a validity argument as presented by Kane (2013).

Another approach to addressing the issue of the quality of the results that are the outcome of the standard setting process is to argue that if the process was well designed, and if the panelists who were selected to participate in the process were appropriate and understood what they were doing, then the results are credible and no further evaluations of quality are needed. This type of argument has been subjected to legal challenge, and it has been found to yield acceptable support for the quality of the standard from a legal perspective (see Phillips (2010) Chapter 8 for a more thorough discussion of legal issues). This is consistent with the philosophical position that setting standards is a judgmental process and if it appears that persons made reasoned judgments, the results can be supported. A counter-position presented by Glass (1978), Shepard (1980, p. 448), and Jaeger (1990, p. 18) is that all processes for determining the point on the reporting score scale are arbitrary or built on arbitrary premises. This view can lead to the conclusion that there are logical flaws in all the procedures, so the process of setting standards should be avoided.

Another view of the evaluation of the standard setting process also takes a critical stance. This view is that participating in a standard setting process involves performing a high-level cognitive task. For the standard to be credible, the panelists who have been selected must be able to perform the task with a high level of fidelity after being given the training that is part of the process. To evaluate the quality of the standard, it is important to obtain evidence that the panelists can perform the required tasks with a high level of competence. This evaluation is usually done by making some assumptions about the expected characteristics of panelists' ratings and checking to determine if the observed results match expectations. If they do not match the expectations, then the location on the reporting scale that is defined by the process is argued to be not credible. For example, Clauser, Clauser, and Hambleton (2014) indicate that panelists should be able to estimate the empirical difficulty of test items, but many are not very good at this type of task. If they are poor at the difficulty estimation task, the panelists are not contributing useful information to a particular type of standard setting task.

Some critics of a standard setting process have argued that the tasks are "too cognitively complex" for panelists to do with any level of confidence (Shepard, Glaser, Linn & Borhnstedt (1993)). In a sense, they are suggesting that performing standard setting tasks is like mentally computing the square root of 534 within five minutes and getting a result accurate to one decimal place. Even with training on a computational algorithm, few could meet this requirement. The same argument is used to criticize standard setting processes. The task is argued to be too complex to be done with the amount of training that is given and the time allotted for the process. The counter-position is that with enough training and experience with the examinee population, the panelists can do the task with sufficient competence for the purposes of the standard (Hambleton & Pitoniak, 2006 p. 441).

There are also statistical procedures for evaluating the outcomes of the standard setting process that parallel ideas from educational testing and psychometric theory. The quality of the standards is evaluated as good if the process results in estimates of the points on the reporting score scale that have small standard errors and that are generalizable across panels. These evaluative criteria are roughly equivalent to concepts from reliability assessment such as internal consistency and stability. However, Shepard (1980, p. 448) notes that there is always error in the selection of a location on the reporting score scale and individuals immediately on either side of the location are "virtually indistinguishable from one another." As a result, she argues, "pass-fail distinctions near the cutoff will have poor validity because a continuum of performance has been 'arbitrarily' dichotomized."

The current thinking about the requirements for the evaluation of the quality of the results of a standard setting process are described in the *Standards for Educational and Psychological Testing* (American Educational Research Association, American Psychological Association and National Council on Measurement in Education (1999, 2014)). These will be discussed in detail in Chapter 9.

1.9 Is a Performance Standard an Instructional Goal?

Before completing this introductory discussion on the definition of a standard, it is useful to separate the idea of a standard as a location on a measurement scale from the idea of a standard as the goal for instruction or learning (Hambleton & Pitoniak, 2006). In the educational literature, there is often a discussion of content standards. Content standards describe the recommended focus of instruction at a particular grade level. Most states in the United States have descriptions of the targets of instruction in the different subject matter areas for each grade. However, the content standards do not usually indicate the amount of the material that needs to be learned to be considered to have reached an acceptable level. Students are not expected to learn everything in the curriculum description before they move to the next grade. Part of the reason for that is that content standards often use open-ended statements like "the student will be able to add integers" without stating the number of digits in the integers or the number of unique integers that must be summed at the same time. It is the standard setting process as considered in this book that is used to determine how much is enough. The amount that students should learn of the material described in a content standard is often called a performance standard (Hambleton & Pitoniak, 2006). Unless specifically stated otherwise, the word "standard" as used in this book refers to a performance standard that is often conceptualized as a location on a hypothetical scale describing the levels of performance, rather than the description of the content that is the target of instruction. Certainly, in many educational contexts, the standard will be related to the target of instruction, but those are considered here as curriculum goals rather than as standards of performance.

1.10 The Goals for This Book

This introductory chapter listed many topics to show the complexity of the processes used to set standards, the issues that have been raised about the ways that the process is implemented, and ways that the quality of the results obtained are judged. The material that follows in this book describes the components of a standard setting process in detail, including formal definitions of each component, and then provides theoretical perspectives about the goals of standard setting and the methods for evaluating the results.

Following the theoretical development, several common methods for setting standards are analyzed using the theoretical concepts with the goal of providing a framework for studying standard setting procedures. A focal point for presenting the examples is to determine whether there are aspects of the standard setting methods that make it preferred over other methods. That is, are all standard setting methods based on equally good conceptual frameworks, or can it be argued that one method is better than another?

The final chapter of this book provides a description of trends in the field of standard setting at the time this book was written. Areas of research and development of the standard setting process and likely areas of improvement are described. This information is presented more as a stimulus for research and development than as an expectation that it is an accurate prediction of future standard setting approaches.

2

A Theory of Standard Setting

2.1 Introduction to a Theory of Standard Setting

Chapter 1 provides an overview of the issues that need to be addressed by a theory of standard setting. The chapter did not provide a comprehensive review of the literature, but it should have given enough details about standard setting processes to show that standard setting is a very complex process. There are many component parts to the process, and the parts interact with each other. Much of the complexity in the standard setting process will be ignored in this chapter so that the basic framework of the standard setting process can be presented in a way that highlights the main features of a theory of standard setting without being distracted by special cases and the unavoidable details needed for practical implementation. Later chapters will expand on the component parts of the process to show how all the complexity and details can be addressed by the proposed theory of standard setting.

A concise statement of the theory of standard setting is that it is a process for translating a conceptual standard defined in a policy statement to a location on a reporting score scale (cut score). Individuals who obtain scores that exceed the standard set on the reporting score scale are inferred to have the characteristics implied by the conceptual definition of the standard. The most critical part of this concise statement of the theory is that what is typically called "standard setting" in the measurement literature is a process of *translation* from policy to a location on a reporting score scale. Persons who participate in a standard setting process are not really setting a standard—the standard is already given in the policy definition. The participants in "standard setting" are translating policy to a location on a reporting score scale. The nuances of this view of standard setting are presented in this chapter. The complexities that are involved in each component part of the standard setting process are dealt with in later chapters.

In subsequent chapters of this book, the term "standard setting process" will be used to refer to an integrated set of activities from the development of policy to the final approval of the location on the reporting score scale and descriptive statements about the meaning of the approved standard. Standard setting process is the most inclusive term used in this book.

DOI: 10.1201/9780429156410-2

"Standard setting method" is used to refer to procedures used to collect data/observations that will be used to estimate the location (cut score) on the reporting score scale. The standard setting method is a part of the standard setting process. The "kernel" of the standard setting method is the cognitive task that is specified by the standard setting method for the purpose of collecting data/observations that are used to estimate the location (cut score) on the reporting score scale. The standard setting method includes more than the kernel. It includes the estimation methodology and other aspects of the process such as the feedback provided to individuals involved in the translation of policy to the location on the reporting score scale. In some cases, more than one kernel is included in the standard setting method—different cognitive tasks may be used at different points in the standard setting process to collect data to support the translation from policy definition to location on the reporting score scale.

2.2 Basic Concepts

2.2.1 Agency

The theory presented here is limited to the standards that are called for by an "authority," as described in the definitions provided in Chapter 1. That is, some professional organization, government agency, school district, etc. determines the need for a standard of performance and develops a policy that calls for the development of the standard. In this book, the term "agency" is used to encompass all the possible groups/organizations that call for a standard. When considered as a general abstract entity, the agency will be represented by the symbol Ag. In this chapter, the agency will be treated as if it were a single entity. In later chapters, it is acknowledged that the agency is likely made up of multiple persons who may not be in total agreement about the desired level for a standard. The variation of perspectives within an agency may influence the final level that is approved for a standard. In its simplest form, the theory of standard setting is

$$\mathbf{S} = f(Ag), \tag{2.1}$$

where **S** is the conceptual standard set by the agency. The convention for mathematical symbols of using bold to represent vectors or matrices is used here to indicate that the conceptualization of a standard has more than one part. It includes at least an implied continuum and a location on that continuum. These concepts will be elaborated later in this chapter.

Equation 2.1 indicates that **S** is a function of the agency. However, **S**, as given in Equation 2.1, is not readily usable for describing capabilities of individuals or for decision-making. It is an unobserved mental conceptualization of the standard by members of the agency. The conceptualization by the agency is typically presented as a word, phrase, or a short paragraph. However, the description of the standard by brief text does not capture the nuances of the conceptualization by the agency. It is a summary or placeholder for the complete, but unobserved, concept behind the standard.

2.2.2 Policy Definition

To make this conceptual standard into something that can be used in a practical way, the agency communicates the details of its thinking to users at some level of detail. The communication about the standard and its meaning is given in a written document that is labeled here as the "policy definition" of the standard. When used in an abstract, technical context, the policy definition is represented by the symbol $\mathbf{D_P}$. As with the conceptual standard **S**, policy definition is in bold indicating that it has more than one component. The following is an example of a policy definition for the conceptual standard labeled "Proficient" that was produced by the National Assessment Governing Board (NAGB), the agency, for use with the National Assessment of Educational Progress.

> Solid academic performance for each grade assessed. Students reaching this level have demonstrated competency over challenging subject matter, including subject-matter knowledge, application of such knowledge to real-world situations, and analytical skills appropriate to the subject matter.
>
> **(NAGB, 1995)**

Policy definitions vary substantially from agency to agency. Some are very specific, and others are very general and possibly ambiguous. However, most of them contain or imply three important ideas. The first is that there is a continuum of performance, sometimes called a construct, and persons vary on their locations on the continuum. The NAGB policy definition implies a continuum by considering "students reaching this level." Other materials from NAGB elaborate on the concept of the continuum generally referring to what schoolchildren know and can do in a subject matter area at their current point in their education.

The second idea is the intended standard on the continuum. For the policy definition given by the NAGB, that is implied by the "solid academic performance … over challenging subject matter." This description of the standard indicates that NAGB is not conceptualizing a standard that is minimum competency, but a standard that is substantially higher on the continuum. The third idea is the use to which the standard would be put. This is important because it indicates the level of accuracy that is needed. NAGB does not include a description of use in its policy description, but other policy documents clearly indicate that the standards will be used for reporting at the state and large, urban district level and not at the student, classroom, or school level. No decisions about individuals will be made from the comparison to the standard.

This fine-grained decomposition of the policy definition can be summarized by the following equation:

$$\mathbf{D_P} \in \left[\text{Construct, Level, Use}\right] = f_P\left(\mathbf{S} \middle| Ag\right), \tag{2.2}$$

where f_P is a function of the standard conceptualized by the agency. This equation shows the component parts of the policy definition using words because the agency typically presents these ideas in conceptual terms rather than using the formal mathematical terms that are typically used by measurement specialists. When a practical standard setting process is implemented, the conceptual descriptions of the components need to be translated into formal measurement concepts. The construct will be represented by the reporting score scale from a measurement instrument. A general symbol for the location on the score scale that represents the construct is ζ. The location on this scale for an individual person, i, is represented by ζ_i. If the reporting score scale is developed using item response theory (IRT), θ is substituted for ζ to indicate that the scale has the typical properties of those developed with IRT models. If the scale is based on summed scores, then τ is used to reference true-score theory as the basis for the scale.

Once a formal measurement scale has been defined for the construct given in $\mathbf{D_P}$, the location of the standard intended by the agency can be considered. The location of the intended standard on the scale implied by the construct is represented by ζ_{Ag}. It can be similarly indicated on the IRT-based scale by θ_{Ag} and τ_{Ag} on the true-score scale.

The "use" part of $\mathbf{D_P}$ describes the inference or inferences that are intended based on the use of the standard. For example, being above ζ_{Ag} on a measure of achievement in high school mathematics may be used to infer that the student will likely be successful in an entry-level course at the college level. The same standard on the same construct might also be used to infer that the student can be successful in an entry-level job after high school. Because standards can have more than one use, the "use" part of Equation 2.2 is represented by $\mathbf{U} = [u_1, u_2, \ldots, u_k]$ with each u representing different uses. The components of $\mathbf{U}$ are important when constructing the validity argument to show that the standard functions as intended. The validity argument is the logical connection of evidence to show that the desired inferences from the standard are supported by evidence.

2.2.3 Content-Specific Definition and Test Specifications

A formal representation of the standard setting process can be represented using the symbols for each of the component parts of $\mathbf{D_P}$. The specification of a standard for operational use requires both the definition of the continuum and the specification of a location on the continuum. Equation 2.3 gives the symbolic representation of the connection between conceptualization by the agency and the location of the standard on the construct:

$$\left[\zeta, \zeta_{Ag}\right] = f_{\zeta}\left(\mathbf{D_P} \middle| Ag, \mathbf{S}\right), \tag{2.3}$$

where f_{ζ} is the function that converts the written description of policy to the location on the continuum implied by the policy definition. This equation indicates that the policy definition $\mathbf{D_P}$ is conditional on the agency and its conceptualization of the standard, and the result of the standard setting process is a location, ζ_{Ag}, on the construct ζ. Note that the uses for the standard, $\mathbf{U}$, are contained in $\mathbf{D_P}$.

It would be ideal if it were possible to translate directly from $\mathbf{D_P}$ to ζ_{Ag}. That seldom, if ever, happens in practice. $\mathbf{D_P}$ is intended to define policy in broad strokes and to communicate to an audience of individuals using or being affected by the policy. $\mathbf{D_P}$ is not written at the level of detail needed to define the construct of the progression of skills and knowledge along the construct. Additional activities are needed to gain the detail that is needed to do the practical work identifying a location on a reporting score scale for a test that is the operational definition of the construct.

The usual first step in providing the needed detail is to translate $\mathbf{D_P}$ into the language of the construct on which the standard is located. This translation is labeled the content-specific definition and is represented as $\mathbf{D_C}$, where $\mathbf{C}$ indicates the content domain or trait that is implied by $\mathbf{D_P}$. As with $\mathbf{D_P}$, $\mathbf{D_C}$ uses "bold" to indicate that the content-specific definition has more than one part. One part is the definition of the trait or domain that defines the construct. A second part is a description of the characteristics of persons who are above the standard on the trait or the skills and knowledge included in the domain. For example, for a licensure/certification examination, the $\mathbf{D_P}$ may indicate that a person must be competent to take on entry-level duties within a profession, but without stating explicitly what those duties are. Then, a job analysis is performed to determine the details of the skills and knowledge that are needed to comply with the $\mathbf{D_P}$. The results of the job analysis provide both a description of the domain, $\mathbf{C}$, and a description of an acceptable candidate. The domain component of $\mathbf{D_C}$ provides the information needed to develop the test specifications for the test that will operationalize the measurement of ζ.

The addition of $\mathbf{D_C}$ to the theory of standard setting indicates that the standard setting process includes multiple levels of translation. Each of these levels can add some error to the overall process, so each translation step needs to be carefully checked to show that it accurately reflects $\mathbf{D_P}$ and ultimately $\mathbf{S}$. What was initially expressed as the simple idea of translation of policy into a number on a scale is fairly complex in practice. First, the policy definition is translated to a content-specific definition:

$$\mathbf{D_C} = f_C\left(\mathbf{D_P} \middle| Ag, \mathbf{S}\right), \tag{2.4}$$

where f_C is the function that converts $\mathbf{D_P}$ to $\mathbf{D_C}$. Next, the policy definition and content-specific definition are translated into test specifications:

$$\text{Test Specifications} = f_{TS}\left(\mathbf{D_P}, \mathbf{D_C} \middle| Ag, \mathbf{S}\right), \tag{2.5}$$

where f_{TS} is the function converting $\mathbf{D_P}$ and $\mathbf{D_C}$ to Test Specifications. While it is ideal if the test specifications are developed specifically to support the uses of the **S**, an existing test is often selected rather than custom-designing and custom-constructing a test to support the use of **S**. The constructed or selected test defines the reporting score scale that is the operational definition of ζ.

In some cases, there is a single policy definition and multiple content-specific definitions. For example, the policy definition provided by the NAGB given earlier in this chapter was meant to apply to academic performance in several different subject matter areas. The policy definition was translated independently into content-specific definitions for each of the subject matter areas. Similar situations exist in state assessment programs that include a definition of "proficiency" that is then translated into specific content in each subject matter area. Test specifications are developed for each of the subject matter areas.

Construction of a test to match the requirements given in the test specifications should produce a test that yields scores that are a good representation of the intended construct, ζ, that is implied by $\mathbf{D_P}$ and $\mathbf{D_C}$. The scores from the test are estimates of ζ, represented by $\hat{\zeta}$. The estimates of ζ from the test scores may have substantial error if the test is not well-designed to correspond to the intended construct. When a standard is set on an existing test rather than a custom-designed test, the test was likely developed before the agency wrote the $\mathbf{D_P}$. In such cases, it is important to determine how well the score scale for the existing test represents the intended construct.

The content-specific definition $\mathbf{D_C}$ refers to the skills and knowledge implied by the $\mathbf{D_P}$ for the range above **S**. The location on the construct that specifies the standard, ζ_{Ag}, is conceptually the lowest level of skills and knowledge that meets the requirements of $\mathbf{D_C}$ and that is consistent with $\mathbf{D_P}$. For some approaches to translating $\mathbf{D_P}$ to ζ, the process focuses on the lowest acceptable level of performance rather than the entire range of skills and knowledge implied by $\mathbf{D_C}$. In such cases, a description of the minimum qualifications to be considered above the standard is often produced. That description is represented here as $\mathbf{minD_C}$. For these standard setting processes, $\mathbf{minD_C}$ is the conceptual definition that is the focus of the translation process onto the construct implied by the policy definition:

$$\mathbf{minD_C} = \text{minimum}\left(\mathbf{D_C} \middle| Ag, \mathbf{S}, \mathbf{D_P}\right). \tag{2.6}$$

2.2.4 Kernel of the Standard Setting Process

Standard setting would be very straightforward if persons could be identified who are bilingual in the language of $\mathbf{D_C}$ and in the language of ζ. If such persons could be recruited, they could perform a direct translation of $\mathbf{D}_C$ or $\mathbf{minD}_C$ to ζ. Unfortunately, few such bilingual persons exist, so testing experts have developed methods to help persons who are proficient in the language of $\mathbf{D}_C$ to perform the translation. The conceptual underpinning of each of the methods that have been developed is labeled here as the kernel of the standard setting method. The kernel is the fundamental cognitive task that is performed by the persons (usually called a panel) selected to take part in the translation process. The result of performing the tasks specified by the kernel are responses from the panel (panelists) that can be used to estimate the point on the construct of interest, ζ_{Ag}, that corresponds to $\mathbf{D}_P$ and $\mathbf{D}_C$ (or $\mathbf{minD}_C$). Many kinds of kernels are used as the basis for standard setting processes. Books such as Cizek and Bunch (2007) describe the various kernels (although they do not use that term) and how they are included in standard setting processes.

The kernel of the standard setting process is the specific cognitive task that a panelist is asked to do to facilitate the translation of the policy and content definitions into a point on the reporting score scale for the test used to operationally define the construct implied by the agency calling for the standard. When panelists perform the task specified by the kernel, they generate a matrix of data points that are then used to estimate the point on the reporting score scale that is consistent with the values in the matrix. Most standard setting processes use a statistical methodology to compute the estimate of ζ_{Ag}, but some processes may get the estimate through discussion and a consensus process.

A simple example of a kernel is direct estimation of the point on the reporting score scale by the panelists. Suppose a panel is composed of ten individuals who are well-versed in the policy and the content definitions that describe the standard. They are also very familiar with the content and other characteristics of the test used to represent the construct implied by the policy definition. Because of the high level of competence of the panelists in the languages of policy, content, and the reporting score scale, they can do a direct translation of policy to a point on the score scale. The kernel is "Specify the point on the reporting score scale for the test that most accurately reflects the minimum level of performance implied by the policy and content definitions." The result is a vector containing a score point from each of the panelists. A statistical method, such as computing the mean value of the scores in the vector, can be used to give an estimate of the standard on the score scale that is the translation of the policy.

The abstract representation of the kernel is given in Equation 2.7:

$$\mathbf{R} = K\left(\mathbf{Panel}, \text{Test Specifications}, \mathbf{D}_C, \mathbf{D}_P \middle| Ag\right), \tag{2.7}$$

where $\mathbf{R}$ is the matrix of information provided by the panelists and K represents the kernel that is selected to assist panelists with the translation process as a function. Because there are many different types of kernels for standard setting processes, the particular one selected is represented by a subscript to the K, such as K_{Direct} for direct estimation. Also, the matrix **R** may have a subscript indicating that it was produced by a particular Panel, $\mathbf{R}_{\text{Panel A}}$, or an individual member of a Panel, $\mathbf{R}_i$.

2.2.5 Estimate of the Intended Standard

The estimate of the standard implied by the policy definition, $\hat{\zeta}_{Ag}$, is then given by

$$\hat{\zeta}_{Ag} = f\left(\mathbf{R} \middle| K, \text{Test Specification}, \mathbf{D}_\mathbf{C}, \mathbf{D}_\mathbf{P}, Ag\right). \tag{2.8}$$

Equation 2.8 highlights the fact that the estimate of the intended standard is a function of all the components in the standard setting process. Changing any one of the components will likely change the estimate of the intended standard. Of course, the agency, *Ag*, can change its intentions and the corresponding estimated standard should change accordingly. In most cases in this book, Equation 2.8 will be represented as $\hat{\zeta}_{Ag} = f(\mathbf{R})$ for brevity, but it should always be understood that **R** is the result of applying the kernel selected for the standard setting process to all the components on the right side of Equation 2.7. For the specific example given above using the direct estimation kernel and the mean (expected value, *E*( )) as the statistical procedure for estimation, Equation 2.8 becomes

$$\hat{\tau}_{Ag} = E(\mathbf{R}), \tag{2.9}$$

if the direct estimation is being done on the number-correct score scale implying that the intended construct is defined using true-score theory.

There are many different kernels that are used for standard setting processes. Sometimes, the same label will be given to variations on a process, or the generic term "modified" will be applied to the label for a kernel. The term "modified" is almost meaningless because it is seldom that the context for the application of a kernel is the same. The basic cognitive task for the panelists may remain the same, but the amount of training and types of feedback will likely be different for every standard setting study. For example, the Angoff kernel usually means panelists are asked to indicate the probability of a correct response for a minimally qualified person on each item on the test. But sometimes, the panelists are instructed to give the probability estimates to the nearest .1 or to the nearest .01. Sometimes they are instructed to take the possibility of guessing the correct response into account and sometimes they are not. The precise kernel is the set of instructions given to the panelists before they generate the information that will be used to estimate the point corresponding to $\mathbf{minD_C}$ on the reporting score scale.

Sometimes, more than one kernel is used with a standard setting method. For example, an Angoff-type kernel might be used first and then, after panelists are familiar with the reporting score scale for the test, a direct estimation kernel might be used. There are an unlimited number of possible variations in the way that a standard setting process can be designed. The large number of variations in the standard setting process makes it very difficult to compare the characteristics of one standard setting method with another one and to make recommendations about standard setting processes that generalize beyond the context of a particular standard setting study.

2.2.6 Feedback to Panelists

One other component needs to be added to the theory of standard setting and that is the influence of feedback to panelists on the process of translation. Because panelists typically are not equally proficient with the language of the reporting score scale, the language of $\mathbf{D}_C$, and the language of $\mathbf{D}_P$, the first attempt at translation, even with the support of a kernel to help guide the process, might be considered as a rough draft. Therefore, feedback is often provided after the first estimates of the standard are obtained to help the panelists refine the information they provide for making the translation. There are many types of feedback that can be provided. The most common feedback after the first draft result, usually called the first "round" of a standard setting process, is the distribution of estimated standards from the different panelists. This feedback allows panelists to observe the level of consistency that has been obtained when translating the $\mathbf{D}_C$ and $\mathbf{D}_P$ to the reporting score scale. This type of feedback is equivalent to several people translating text from one language to another and then sharing the first draft of their translations. The feedback allows them to note differences in word choice and the structure of the translation, and possibly see misinterpretations of some specific part of the text. Translators may discuss the differences to come to agreement on the text for an accurate translation.

Other types of feedback can be given as well. It is common to provide some information about the observed difficulty of test items when the kernel that is used involves estimating probabilities or responses to individual items. This is the equivalent of providing formal definitions of words to help with a translation from one language to another. Sometimes the proportion of persons who exceed the estimated standard is provided. This is an example of providing norm-referenced information, as discussed in Chapter 1. The goal of providing feedback should always be to improve the quality of the translation of the $\mathbf{D}_C$ rather than to shift the standard away from what was intended by the agency when it created $\mathbf{D}_C$. Even with translations from one language to another, there are sometimes pressures from feedback to avoid the use of certain words that might be considered offensive or make a statement less forceful than the original author intended.

In most cases, the standard setting method is a recursive process that includes the use of a kernel to acquire information to get an estimate of the standard on the reporting score scale, then give feedback to panelists to allow them to refine the information they provided in response to the requirements of the kernel, and then go through the process again. It is common to have panelists go through the process two or three times to refine the translations and stabilize the overall results. The expectation is that the estimates on the reporting score scale will converge over rounds. The ideal would be that the estimates converge to the standard that is intended by the agency.

The recursive nature of standard setting can be represented in a general form by the following expression:

$$\mathbf{R}_{\mathrm{m+1}} = K_{m+1}\left(\mathbf{F}_{\mathrm{m}}, \text{Test Specifications}, \mathbf{D}_{\mathrm{C}}, \mathbf{D}_{\mathrm{P}} \middle| Ag\right), \tag{2.10}$$

where $\mathbf{F}_{\mathrm{m}}$ is the feedback given after round m,
K_{m+1} is the kernel that is used by the panelists in round $m + 1$,
and $\mathbf{R}_{\mathrm{m+1}}$ is the information obtained from the panelists at the end of round $m + 1$.

The initial round of rating is numbered as round 1 ($m = 1$), so $\mathbf{F}_1$ is the feedback after the initial implementation of the kernel. The kernel has a subscript in Equation 2.10 to indicate that the kernel may change from round to round. The final estimate of the standard on the reporting score scale is computed from $\mathbf{R}_{\mathrm{m+1}}$ after the final round of the process.

2.2.7 Approval of the Estimated Standard

The final step in the standard setting process is typically providing the standard estimated from $\mathbf{R}_{\mathrm{m+1}}$ to the agency to determine if the translation to the point on the reporting score scale is consistent with the standard intended by $\mathbf{D}_{\mathrm{P}}$. Because the agency does not usually have proficiency in the language of the reporting score scale, information is usually provided about the entire standard setting process and the consequences of the use of the recommended standard, such as the proportion of examinees that exceed the cut score on the reporting score scale. The critical elements are whether the steps in the standard setting process are logically consistent, and if the consequences of applying the estimated standard are consistent with the expectations of the agency. Often, the agency has access to other information such as performance of examinees on other exams, or other measures of success. The agency does not know what the "correct" value is on the reporting scores scale, but it can indicate if results do not match expectations. For example, if the expectation is that the standard is for minimum competency and that most examinees will exceed the standard, but results from the standard setting process show that only 20% exceed the standard, the agency will likely indicate that the results do not match what was intended and the standard should be adjusted, or the entire process should be redone. Sometimes, the agency only requires small adjustments to the standard. These adjustments might be made within the range of error of estimate for the point on the reporting score scale, with the idea that if the entire standard setting process were independently replicated, the results would be slightly different. For this reason, standard errors of the estimates of the estimated point on the reporting score scale are often computed along with the estimate itself.

2.3 Documentation of the Standard Setting Process

As indicated numerous times in the description of the component parts of the standard setting process, no two standard setting processes are the same. Therefore, it is important that all the details of the standard setting process be well-documented so that the agency and other interested parties will know exactly what was done and how the process differs from others used elsewhere. There is no ideal bookmark or Angoff process: the same kernel can be applied in many ways and each variation will likely give somewhat different results. Truxillo, Donahue, and Sulzer (1996) also indicate that good documentation helps support the legality of the standard in the employment testing context.

Ultimately, if a well-designed standard setting process has been implemented, including all the components described above, the agency approves a standard that falls within the range that is considered consistent with the $\mathbf{D_P}$. It is this standard that is used in an operational setting using the test that is the operational definition of the construct of interest. This may be the end of the process, or there may be a review of the standard after some period of operational implementation. If the test used as the operational definition of the scale was newly developed, performance by examinees may change with an increase in experience and as information about test content becomes available. As a result, the proportion scoring above the standard may change over time. The agency may review results after a period of implementation to determine if they continue to match the intentions implied by the $\mathbf{D_P}$.

2.4 The Validity Argument

There are two important inferences that are consistent with standard setting processes. The first is that persons vary on the construct that is implied by the $\mathbf{D_P}$. The expectation is that their performance should also vary on a test that is well-designed to measure the construct of interest. The second inference is that those persons who exceed the cut score on the reporting score scale for the test have the characteristics described in $\mathbf{D_P}$ and $\mathbf{D_C}$. Part of the documentation for a standard setting process should be the validity argument that gives support to these inferences (Kane, 1994). Evidence should be provided that the test does indeed measure the construct of interest and that the reported scores provide acceptably accurate estimates of the location of individuals on the construct in the region of the reporting score scale near the cut score. Further, evidence should be provided that those who are estimated as above the cut score have the characteristics defined in $\mathbf{D}_P$ and $\mathbf{D}_C$. Also, those estimated to be below the cut score should have some deficiencies relative to those definitions.

The theory of standard setting presented in this chapter provides a framework for obtaining and providing evidence to support the inferences implied by standard setting. Clear $\mathbf{D}_P$ and $\mathbf{D}_C$ give useful descriptions of the characteristics expected for those estimated to be above the standard. They also provide useful descriptions of the construct that is the focus of the standard setting process. Thorough documentation of the standard setting process provides much of what is needed for the validity argument to support the use of the standard.

2.5 Some Fanciful Examples of Standard Setting

To help make the standard setting process more concrete, two examples are presented to emphasize important aspects of standard setting. These examples are purposely presented as unrealistic and somewhat humorous because they are not meant to be a template to be followed exactly for a standard setting process. They are presented to show the general framework for standard setting without getting too entangled in the details of the process.

2.5.1 How Much Oxygen Is Needed for Healthy Living?

A health organization has concerns about the amount of pollution in the air and the increase of other normal components of the air such as carbon dioxide. This organization, which meets the definition of an "agency" given earlier, calls for a standard for the amount of oxygen intake needed by adults in the United States so that they do not suffer any ill-effects from lack of oxygen.

Because the agency is trying to do things in a thorough and well-supported way, they initially craft a policy definition of sufficient oxygen for an adult in the United States. This forced them to consider the situations to which the standard is meant to apply. Persons need more oxygen when participating in strenuous activity. There is also less oxygen available at high altitude. The airline industry lobbies for a low standard because they are concerned what a high standard might require related to the pressurization of the cabins of airplanes. There is also a need to determine the scale to be used for reporting the amount of oxygen. Should it be the number of oxygen molecules, the number of cubic feet or meters of oxygen available at a standard pressure, or the partial pressure of oxygen relative to other gases?

The agency ultimately decides that the standard they propose should apply to enclosed spaces and should be based on the minimum partial pressure of oxygen given that the atmospheric pressure within the space is between sea level and that at 6,000 feet. The proposed policy definition, $\mathbf{D_P}$, of the standard is as follows:

> Each enclosed space that is regularly inhabited by adult human beings should have sufficient oxygen so that the physiological functions of the adult body, including brain activity, have no negative consequences. The minimum standard for sufficient oxygen is given in terms of the partial pressure of oxygen in the atmosphere in the enclosed space.

Given this $\mathbf{D_P}$, it is now important that a content-specific definition, $\mathbf{D_C}$, be produced. In this case, the agency calls together a panel of experts on human physiology and pulmonary functioning to specify the scale for measuring the partial pressure of oxygen, the individuals to which the standard will apply, and the types of enclosed spaces that should be included. The panel would also consider the kinds of negative consequences that could occur from insufficient oxygen. The result of the work of this panel is a detailed and somewhat technical description of what it means to have sufficient oxygen and the conditions to be covered by the standard.

An outcome of the development of $\mathbf{D_C}$ for the standard may be an explicit statement of the measurement scale that defines the continuum, ζ, on which the standard will be placed. In this case, the panel recommended the "bar" as the unit of measure and the agency accepted the recommendation. Thus, the task of setting the standard is determining how many bars of oxygen is the minimum partial pressure that is needed to meet the requirements of the $\mathbf{D_P}$ and $\mathbf{D_C}$.

The next step in the standard setting process is to determine the method for determining the point on the bar scale for partial pressure of oxygen that is the translation of $\mathbf{D_P}$. In this case, the standard could be determined directly by placing individuals in an enclosed space and manipulating the oxygen level to see when there are negative consequences. However, this would be based on a limited sample of people and conditions. Because there is already substantial research on oxygen requirements from aviation and submarine research, as well as the space program and special medical applications, the agency instead called for a panel to review the literature on oxygen requirements for human activities and then make a consensus recommendation for the standard. The major important task in gaining this information is selecting a panel with sufficient expertise and representativeness to produce a credible recommendation for the value on the oxygen partial pressure scale.

The panel will have to be carefully instructed about the intent of $\mathbf{D_P}$ and the elaboration contained in the $\mathbf{D_C}$. The panel will also need to determine a way to aggregate information from many research studies that might not be consistent with each other. A good standard setting plan would give guidance about how this aggregation would be done. In this case, higher value would be given to research studies that focused on representative samples of the adult population rather than athletes or persons in the space program. The resulting standard is intended to apply to daily life instead of highly stressful physical or mental activities.

The panel's task is to make a recommendation of the point on the oxygen partial pressure scale, $\hat{\zeta}_{Ag}$, that is consistent with $\mathbf{D_P}$ and $\mathbf{D_C}$. The panel may also give information about their level of agreement with the result, and they should provide an evaluation of the overall process. All information is provided back to the agency calling for the standard for final approval.

The agency reviews all material and determines either to approve the standard as recommended or to modify the standard based on other information that was not provided to the panel. For example, the agency may have access to information about the financial cost of implementing the intended standard. They may have been informed about the number of sites that would be in compliance with the standard. There might be information from other countries. If the agency does adjust the standard, then care must be taken to ensure that $\mathbf{D_P}$ and $\mathbf{D_C}$ are consistent with the final standard. If not, these definitions must be adjusted to match the final result.

This example is fairly simple because there is an existing scale for defining the continuum of available oxygen and there is a large body of research on oxygen use by the human body. But, even for this simple case, there are many complicating factors and producing such a standard would be a nontrivial activity.

2.5.2 Understanding of the Functioning of Passenger Automobiles

A state agency that is responsible for giving out driver's licenses has noticed that some persons who apply for a license know very little about the cars they are driving. For example, they do not know how to adjust the mirrors or to set the parking brake. Some of this lack of knowledge is related to safety. The current requirements for a driver's license are mainly about knowing the published rules for driving and how to drive the car, but there is little knowledge required about how the car functions. Therefore, the agency calls for a standard of knowledge and skills about the functioning of the car that they are driving that should be included as part of the requirements for getting a driver's license.

As a starting point for developing the standard, the agency proposes the following policy definition, $\mathbf{D_P}$, for the standard:

> It is important that a person who is given a license to drive a car know enough about the vehicle they are driving to be able to use it safely. This

includes knowing the basic principles about how the car works along with knowing how to use critical components of the car such as the controls for adjusting mirrors, seats, cruise control, lights, etc. They should also know how to properly use seat belts, lock and unlock doors, access the engine, fill the gas tank, etc. These are not a complete list of the required knowledge and skills. The licensed driver should have sufficient knowledge of the working of the vehicle to drive it safely.

Following the development of $\mathbf{D}_P$, the agency called together a group of individuals from auto manufacturers, police departments, and safety advocacy groups to develop a content definition, $\mathbf{D}_C$, for the standard for automobile knowledge and skills. This group immediately noted that automobiles are different in the ways that controls are organized, so they needed to determine common components across vehicles as the focus for the standard.

From the work of this group with varied interests and backgrounds came a $\mathbf{D}_C$ of the skills and knowledge needed to safely operate a car. This $\mathbf{D}_C$ defined a domain of knowledge and skills, and it implied a continuum of knowledge and skills, ζ, describing drivers who are unable to even enter and start the vehicle to those with an in-depth understanding of all its component parts.

The agency calling for the standard then contracted with a test development company to develop a knowledge and skills test that was consistent with the definitions. The scores on this test became the operational definition of the continuum on which the standard would be set. The expectation was that the test would be sufficiently reliable to make accurate estimates of individuals' locations on the continuum. The agency also arranged for multiple forms of the test to be developed that would be parallel in content and psychometric characteristics. The expectation was that the standard would be fairly low on the continuum because persons would be able to drive a car without being an automotive engineer, so the precision of the scores on the test should be targeted at that level.

The agency then needed to determine who was qualified to convert the $\mathbf{D}_P$ and $\mathbf{D}_C$ to a point, ζ_{Ag}, on the continuum defined by the test scores. Certainly, the individuals should be able to safely drive a car. They should also be familiar with the skills of novice drivers and the variety of features on cars that were commonly driven. The agency determined that highly qualified driving instructors should be included on the panel. They also included representatives of the police department that were actively involved in enforcing traffic laws or responding to accidents involving cars. They decided that representatives from auto safety organizations should be included.

The agency then hired an organization to implement the standard setting process. This was done through an open bidding competition and possible vendors were allowed to recommend the standard setting procedure that they thought was best. Because there was no data on the test at the time of first implementation, the selected vendor recommended a modified Angoff method because a passing score on the number-correct score scale could be defined without extensive test analysis. However, to check the process, they recommended that the process be conducted with two separate test forms so that the stability of the passing score could be determined. The panel would also be divided into two groups that would work independently so that the estimated standards could be compared.

The standard setting process would take two days, with the first day devoted to training the panel on the policy and content definitions, and on the standard setting process. The second day would be devoted to two rounds of standard setting with feedback between rounds and an evaluation at the end of the process. After the process was completed, the vendor would prepare a report that gave the details of the process and the evaluation of the process by the panel, as well as technical results about the level of agreement of the two groups from the panel and the agreement between the estimated standards set on different test forms. From all this information, the vendor would present a recommended point, $\hat{\tau}_{Ag}$, on the score scale for the test as the translation of policy.

The agency reviewed the results after receiving the report from the vendor and determined that the recommended standard was lower than they expected. They chose to adjust the standard upward and carefully documented the reasons, and then reviewed the $\mathbf{D}_P$ and $\mathbf{D}_C$ to make sure they were consistent with the standard. As a final step, the agency provisionally implemented the standard for six months to determine the passing rates for individuals and the relationship between scores on the test and accident rates for drivers. Although the test scores were found to have a moderate correlation with the likelihood of having an accident, the overall proportion of accidents was very low, so the agency decided to move the standard back to the point recommended by the panelists. All the deliberations were carefully documented to support the results should there be any legal challenges to the use of the standard.

2.6 Summary and Organization of Subsequent Chapters

This chapter has provided the basic structure of the theory of standard setting. It lays out the basic components both conceptually and in a more abstract form. It also provides the connection to the validity argument for a standard. Finally, it gives two fanciful examples to make the concepts more concrete. However, real standard setting processes are much more complex than the idealized version presented in this chapter. Subsequent chapters will delve into the details of the standard setting process with the goals of showing how the theory applies in real settings and how the theory helps identify weaknesses in typical standard setting studies.

After describing the details of each component in a standard setting process, the concept of validity argument is covered in more detail along with the documentation needed to highlight the unique features of any standard setting process. The final chapter describes recent trends in research and development that could improve the overall approach to standard setting.

3

Agency, Policy Definitions, Content-Specific Definitions and Minimum Qualifications

3.1 The Agency and Its Role in Setting Performance Standards

The theoretical position taken in Chapter 2 is that formal standards do not come into existence on their own; they are called for by some authoritative body that is given the general term "agency" and is represented by the symbol *Ag* in this book. There are many kinds of authoritative bodies that are included under the abstract term "agency," and as a result, there are many different ways that the agency is involved in the standard setting process. At the very least, the agency calls for the existence of a standard, and in doing so, it typically creates a policy that includes something about the uses of the standard.

The agencies that call for a standard vary in many ways. It is helpful for the development of a theory of standard setting to consider three of the many dimensions on which they vary. These three are: (1) involvement, (2) expertise, and (3) heterogeneity. Each of these three dimensions will be defined and described, followed by some examples of agencies with different combinations of the dimensions.

3.1.1 Involvement

The level of involvement refers to how much the agency is actively participating in the design, application, and approval of the standard setting process. At one extreme is an agency that calls for the existence of a standard and then delegates all the design, application, and even the approval of the final estimated standard to contractors or other organizations. This low level of involvement often occurs when the agency is a state legislature, and those who called for the standard are no longer in office when the estimated standard has been obtained.

In most cases, the agency has a level of involvement that includes at least producing a policy definition, $\mathbf{D}_P$, for the standard and reviewing and approving the final estimated standard, $\hat{\zeta}_{Ag}$, as being consistent with $\mathbf{D}_P$.

DOI: 10.1201/9780429156410-3

In other cases, the agency actively reviews proposals for the standard setting process and makes policy decisions about desired characteristics of the panel involved in translating the standard implied by $\mathbf{D_P}$ to the reporting score scale for the test that is the operational definition of the construct also implied by $\mathbf{D_P}$. Involvement can be thought of as a continuum from minimal involvement to careful review of all aspects of the standard setting process.

3.1.2 Expertise

Expertise refers to the amount of background that the agency has related to the construct on which the standard will be set. Especially in the case of certification and licensure tests, the agency is composed of individuals who are highly competent in the area that is the focus of the standard. However, in other cases, the agency may call for a standard on a construct even though the members of the agency are not skilled in that area. This happens frequently when a state board of education calls for a proficiency standard for achievement in science or mathematics when they do not have high levels of expertise in those areas.

It is often the case that members of the agency will have expertise in different areas, and collectively they include all the areas contained in $\mathbf{D_P}$. In other cases, the agency will call on others with the desired expertise to craft the content-specific definition, $\mathbf{D_C}$. The importance of expertise in the standard setting process is the capability of the agency to judge whether $\mathbf{D_C}$ accurately reflects their intentions and whether the estimated standard is consistent with all the information that is considered in the process.

Expertise relates to other features of the standard setting process besides the construct that is assessed by the test. It also relates to knowledge about the examinee population and the uses for the standard. If the standard is for achievement of third grade students, it would be helpful if the agency had expertise about the capabilities of third grade students. If the standard is for English language competency for persons attending graduate school in business, it is useful to understand the context for use of the language.

Overall, agencies can be thought to vary along a continuum from very high expertise with members directly involved in using the construct on which the standard will be set to very low direct knowledge and skills on the construct. In the latter case, the agency is a policy group that is dealing with more general issues, such as the quality of education in the country, rather than the specific skills and knowledge for a narrowly defined profession.

3.1.3 Heterogeneity

Heterogeneity refers to the diversity of backgrounds and positions taken by members of the agency. The members of many agencies that are related to government programs are purposely selected to represent different constituencies. Such agencies will likely have high heterogeneity. Agencies that are related to a professional specialty area or a specific knowledge area will likely have low heterogeneity. The heterogeneity of an agency will likely influence the specificity of $\mathbf{D_P}$. If the members of an agency have diverse viewpoints, it is likely that any $\mathbf{D_P}$ will have to be written in general terms so that all members of the agency can agree to it. The amount of heterogeneity will also influence the range of locations on the reporting score scale that the agency will consider acceptable for the estimated standard.

3.1.4 Categorization of Agencies

It is useful to think of agencies as categorized onto one of eight cells in a cube defined by the combination of being high or low on each of the three dimensions of the cube. For example, an agency could be high on heterogeneity, low on involvement, and low on expertise. This might be a policy agency such as a state legislature that calls for a standard on a highly technical area. Another agency might be high on heterogeneity, high on involvement, and low on expertise. In this case the agency is representative of many constituencies, but it takes an active role in developing and approving the standard setting process. It also may approve the estimated standard. A third agency might be low on heterogeneity, low on involvement, and high on expertise. This agency might be calling for a standard on a narrow specialty area with members who are all highly competent in that area. However, because the members of the agency are highly active in their own professions, they delegate the standard setting process to a contractor and have minimal involvement in the details.

The standard setting process for each of the example agencies will function in a unique way. The $\mathbf{D_P}$ will be written with various levels of detail, and approval processes for the estimated standard will be different. Several examples of agencies are provided below to show the variation in types of agencies and how they fit into this descriptive structure.

3.1.4.1 State of Michigan Legislature

In October of 2016, the legislature of the State of Michigan passed legislation (Act No. 306 Public Acts of 2016) that required schools to administer reading tests to students in Grades K to 3 to determine if they were likely to reach the Proficient level on the state assessment when tested in Grade 3. The legislation requires that the schools identify the students who have a "reading deficiency" and provide them an instructional intervention to improve their reading capability.

> Select 1 valid and reliable screening, formative, and diagnostic reading assessment system from the assessment systems approved by the department under subsection (1)(a). ... For any pupil in grades K to 3 who exhibits a reading deficiency at any time, based upon the reading

> assessment system selected and used under subdivision (a), provide an individual reading improvement plan for the pupil within 30 days after the identification of the reading deficiency.
>
> **(p. 2)**

A definition of "reading deficiency" is given later in the legislation. "'Reading deficiency' means scoring below grade level or being determined to be at risk of reading failure based on a screening assessment, diagnostic assessment, standardized summative assessment, or progress monitoring" (p. 7).

In this example, the Michigan Legislature is the agency who calls for the existence of the standard and the legislation provides the $\mathbf{D_P}$ for the standard. The $\mathbf{D_P}$ indicates that there will be a point on the score scale on the reading assessments that have been approved by the state's Department of Education, which identifies students who are deficient on the reading continuum, and it also indicates the purpose for the standard—identifying students who need additional assistance to reach state educational standards for reading. The $\mathbf{D_P}$ includes all the component parts given in Equation 2.2 in Chapter 2.

In this case, the agency (state legislature) is high on heterogeneity and low on involvement and expertise. The agency has wide representation from every part of the state and the elected representatives have a variety of backgrounds. It is unlikely that more than a few of the legislatures have a background in reading assessment or reading instruction, so the expertise level is low. Further, once the legislation was passed, the implementation was delegated to Department of Education and individual school districts. The state legislature would not be involved in determining how the standard would be set on a reporting score scale, nor would they approve final standards. It was likely that some of the legislators who were involved in crafting the policy would no longer be in office when the policy was implemented. In this case, the agency has delegated almost all development of the standard setting process to others. However, if the standards were set too low or too high, it is likely that the agency would notice and indicate that the result was not what they intended. That implies that there was an intended standard in the collective mind of the legislature.

3.1.4.2 National Assessment Governing Board

The National Assessment Governing Board (NAGB) is the organization that is responsible for the oversight of the National Assessment of Education Progress (NAEP), a large-scale survey of educational achievement in the United States. In 1990, NAGB decided that it would be useful to report results on NAEP related to standards of performance (Vinovskis, 1998). Since that time NAGB has been actively working to refine the standard setting process used to set standards on the reporting score scales for the different content area versions of the NAEP. As defined in Chapter 2, NAGB is the agency that called for the standards on NAEP. In 2018, NAGB adopted a document that describes the standard setting process that it uses to set the standards on the reporting score scale for NAEP (NAGB, 2018).

As for all standard setting processes, NAGB has a $\mathbf{D_P}$ for the standards that it proposes. For NAEP, NAGB proposes three levels of standards.

> **NAEP Basic**
>
> This level denotes partial mastery of prerequisite knowledge and skills that are fundamental for performance at the NAEP Proficient level.
>
> **NAEP Proficient**
>
> This level represents solid academic performance for each NAEP assessment. Students reaching this level have demonstrated competency over challenging subject matter, including subject-matter knowledge, application of such knowledge to real world situations, and analytical skills appropriate to the subject matter.
>
> **NAEP Advanced**
>
> This level signifies superior performance beyond NAEP Proficient.
>
> (p. 2)

In addition to these $\mathbf{D_P}$s, NAGB produces $\mathbf{D_C}$s for each area that is tested and approves a design document for the standard setting process. On completion of the process, NAGB reviews and approves the points in the reporting score scale that correspond to the $\mathbf{D_P}$s. This example of an agency is involved in every step of the standard setting process. Using the classification scheme suggested in this chapter, NAGB is heterogeneous (members represent many different constituencies), highly involved, and low expertise. Related to expertise, NAGB does have some members who are expert in some of the subject-matter areas that are tested, but most members are not. NAGB contracts with content experts to develop the $\mathbf{D_C}$s and then review them to determine if they match intentions expressed in $\mathbf{D_P}$s.

3.1.4.3 National Council of State Boards of Nursing

The National Council of State Boards of Nursing (NCSBN) is a not-for-profit organization whose members include the nursing regulatory bodies from the states and territories of the United States, including the District of Columbia. NCSBN has an 11-person board of directors which, for the purposes of this chapter, is considered the agency that calls for a standard. One of the services provided by NCSBN is the licensing of nurses for practice. The $\mathbf{D_P}$ for the standard for licensure is provided in NCSBN (2011).

> The purpose of a professional license is to protect the public from harm by setting minimal qualifications and competencies for safe entry-level practitioners. Nursing is regulated because it is one of the health professions that poses a risk of harm to the public if practiced by someone who is unprepared and/or incompetent.
>
> The general public may not have sufficient information and experience to identify an unqualified health care provider, and is vulnerable to unsafe and incompetent practitioners. A license issued by a governmental entity (e.g., the state board of nursing [BON]) provides assurance to the public that the nurse has met predetermined standards.
>
> **(p. 3)**

The actual scope of practice that is required is determined through an extensive review of entry-level nursing positions. The standard that is set is designed to determine if a candidate has "the amount of nursing ability currently required to practice competently at the entry level."

In this case, the agency is homogeneous, consisting of persons active in the nursing profession – expert because of the selection process for Board membership, and involved because they approve the standard that is set on the reporting score scale and review the entire standard setting process. The agency is also involved in determining the specific content that will be on the test that operationally defines the continuum implied by the policy definition.

3.1.5 Influence of Agency Characteristics on Standard Setting

The examples that have been presented show that the agency that calls for a standard may have quite different characteristics. These different characteristics will have an influence on the standard setting process to the extent that they impact the $\mathbf{D_P}$ and $\mathbf{D_C}$ for the standard, the definition of the continuum on which the standard is set, and on the approval process for the translation of the intended standard onto the reporting score scale. The ideal case is for the agency to produce a very precise and clearly stated $\mathbf{D_P}$ that can be understood by all who are involved in the standard setting process. Such a $\mathbf{D_P}$ is more likely to be produced when the agency has low heterogeneity and high expertise. It is also more likely when the agency is directly involved in the standard setting process because the agency will understand how the $\mathbf{D_P}$ will be used.

When the agency has high heterogeneity and low expertise, it is less likely that it will produce a $\mathbf{D_P}$ that results in a unique translation to the reporting score scale. The ambiguity of the $\mathbf{D_P}$ can be avoided if the agency accepts its lack of expertise and understands the differences of its members and involves external experts to help in the drafting of the definitions and defining the continuum on which the standard is set. The example of the NAGB given above is of an agency that has brought many different types of experts into the process to help produce a coherent set of definitions.

To present these issues in a more formal way, the agency, *Ag*, can be thought of as a vector of entries rather than a single entity. When that difference is being emphasized, it will be represented in bold as **Ag** = [Ag_1, Ag_2, Ag_3, …, Ag_i, …, Ag_n] with the individual members as elements of the vector. Further, each of the members of the **Ag** has an intended standard that may be different depending on the heterogeneity of the **Ag**: [ζ_{Ag1}, ζ_{Ag2}, ζ_{Ag3}, …, ζ_{Agi}, …, ζ_{Agn}]. A theoretical indicator of the heterogeneity of an **Ag** could be the standard deviation of the intended standards, and the intended standard that is the focus of the translation process could be the median or mean of the intended standards.

Of course, these intended standards are never known in practice and there is no way to compute measures of heterogeneity or central tendency. However, it is likely that for an agency with high heterogeneity to produce a $\mathbf{D_P}$ that all members endorse, it will have to be written in a very general way. If there is high agreement on the intended standard, the $\mathbf{D_P}$ will likely be much more detailed and precise. Those who implement the standard setting process prefer a detailed and precise $\mathbf{D_P}$.

The amount of heterogeneity in the agency will also influence the approval of the intended standard. The range of intended standards within the agency specifies the range of possible acceptable translations of the intended standards onto the reporting score scale. If the estimated standard falls within the range, it will likely be approved as the final result of the standard setting process. However, the agency may adjust the estimated standard if they judge it to fall outside of the acceptable range.

No research has been identified that investigates the influence of the characteristics of the agency on the standard setting process. It would seem likely that the standard setting process would run more smoothly if the agency had low heterogeneity and high expertise. If that is the case, the $\mathbf{D_P}$ will likely be very clear and accurately describe the desired capabilities of persons being evaluated.

3.2 Policy Definition ($\mathbf{D_P}$)

The agency that calls for the existence of a standard communicates that need for the standard in a statement of policy called the policy definition, $\mathbf{D_P}$, in this book. The $\mathbf{D_P}$ is a critical part of the standard setting process because it is the expression of the intended standard, ζ_{Ag}, by the agency. The $\mathbf{D_P}$ is the written representation of the intended standard, and the other parts of the standard setting process are designed to accurately translate the $\mathbf{D_P}$ into a point on the reporting score scale. In early work on standard setting, the importance of the $\mathbf{D_P}$ was not appreciated. Nedelsky (1954) uses the terms "adequate achievement" or "satisfactory achievement" to describe the standard to be set for grading students in an instructional setting. These labels were the closest he came to providing a $\mathbf{D_P}$. He did indicate that his proposed standard setting procedure might not work very well if the instructors involved in the process did not have good agreement about the meanings of the terms.

The theory of standard setting proposed here indicates that a good $\mathbf{D_P}$ will contain information related to three components of the standard setting process. First, the $\mathbf{D_P}$ should clearly indicate the construct (continuum) on which individuals are expected to differ, and on which the intended standard is set. In the nursing example provided in Section 3.1.4.3, the construct was defined by "the amount of nursing ability currently required to practice competently at the entry level." The $\mathbf{D_P}$ for nursing licensure implies that there is a continuum of knowledge and skills labeled "nursing ability," and that candidates for licensure vary on this ability. Often the construct implied by the $\mathbf{D_P}$ is elaborated in other supporting materials so that there is a clear understanding of what is required of a qualified candidate. It is helpful if the agency is very explicit about the characteristics of the continuum that it has in mind.

The second component of a good $\mathbf{D_P}$ is a description of the "level" of the intended standard. In the $\mathbf{D_P}$ for nursing, the level is defined by "required to practice competently at the entry level." The description of level needs to communicate the intended standard to the persons who will be involved in the translation of the intended standard to the reporting score scale. For the nursing example, the persons involved in the translation are practicing nurses, and the expectation is that they will know the requirements of an entry-level position and what is needed to be competent in that position. Of course, elaboration and precise language when defining the level will help the translation process.

The third component of a good $\mathbf{D_P}$ is a description of how the standard will be used in practice. This is important because it indicates the level of accuracy that is needed when estimating the location on the reporting score scale that corresponds to the intended standard. If the standard will be used informally, such as recommended hours of sleep for the general public, then the translation process can tolerate variation and ambiguity. However, if the standard is used to make important decisions, such as whether a person can be hired in a nursing position, then care needs to be taken to make sure the standard accurately reflects the intentions of the agency. There may be legal challenges to the standard and the agency may need detailed documentation to show that the translation process is accurate and replicable.

Thus, as described in Chapter 2, the $\mathbf{D_P}$ can be summarized as a vector of information, $\mathbf{D_P} \in$ [Construct, Level, Use], with three components. It is important that the components accurately communicate to all those involved in the standard setting process. Further, this vector corresponds to some abstract aspects of the theory: [ζ, ζ_{Ag}, Use]. Note that Use cannot be represented in an abstract way because of its connection to activities outside of the theory.

3.3 Content-Specific Definition (D_C)

The D_P is usually written in general terms either because it is designed to communicate to a general audience or because the agency does not have expertise on the construct implied by the standard to craft a more detailed description. In either case, the first step in the translation of the D_P to a point on the reporting score scale is to translate it into a description of the specific content and skills needed by a person who exceeds the standard. This is called the "content-specific definition" in this book, and is represented by D_C.

Prior to the 1990s, creating a D_C was not a usual part of the standard setting process. Panelists translated the D_P to the reporting score scale without this intermediate step. Hanshe (1998) provides several examples of the standard setting processes used prior to that date. Panelists sometimes produced D_Cs following the standard setting process to help interpret the practical meaning of the standard. Sometimes these "after-the-fact" definitions were seen as inconsistent with the actual performance of examinees. Therefore, Mills and Jaeger (1998) recommended that panelists draft the D_C before performing the ratings required by the kernel for the standard setting process. Later, the NAGB (2018) required that the D_C (they labeled these as achievement level descriptions (ALDs)) be produced at the same time as the design for the test used to operationally define the construct on which the standard is set. This allowed NAGB to approve the D_C before panelists used it in the standard setting process. An example of an ALD produced for the Proficient level of the NAEP for writing is given below.

> Twelfth grade writing at the Proficient level should demonstrate in the allowed time period a competent response to the topic that addresses a specific audience and that serves a clear purpose. Relevant examples, reasons, or anecdotes support the topic and consider the audience. A clear organization provides a focus on the topic and underscores the purpose. Sentences are varied to maintain a reader's interest and express the relationships among ideas. Specific word choices reinforce the purpose of the writing and are appropriate for the topic and audience. The tone of the writing matches the purpose and audience, which may be formal or casual, distant or friendly, depending on the situation described in the writing assignment. Writing at the Proficient level reflects good control of grammar, usage, and mechanics with few if any errors that distract or impede understanding.
>
> **(Bourque & Webb, 2011; p. 6)**

Current recommendations related to the D_C are that they should be produced early in the standard setting process: "The PLDs [Performance Level Descriptions = Content-Specific Definitions] should be written before the standard-setting process, ideally at the beginning of the test development process so that they are a fully integrated aspect of test design and development" (Pitoniak & Cizek, 2016). Egan, Schneider, and Ferrara (2012) give an extensive discussion on the process for producing what they label as "Range PLDs," which are essentially the same as the $\mathbf{D_C}$, as that term is used in this book. They suggest a "committee of educators and cognitive scientists" should write the $\mathbf{D_C}$ to take advantage of their expertise. But the definitions need to be connected to the content included in the test that will operationally define the construct where the standard will be located.

A key concept from the perspective of the theory of standard setting presented here is that the agency has an intended standard, and the intended standard is described in the $\mathbf{D_P}$. The $\mathbf{D_P}$ is translated into a $\mathbf{D_C}$ that panelists will use to guide their judgments as they perform the tasks specified in the standard setting kernel. The $\mathbf{D_C}$ should be of sufficient detail to help panelists with their task, and the agency should approve it to ensure that it captures the intended standard and is consistent with the $\mathbf{D_P}$. At each step in the standard setting process, there should be checks in place so that the intended meaning will be properly translated.

3.4 Minimum Qualifications Definition ($\mathbf{minD_C}$)

A critical component of many standard setting processes is the definition of the characteristics of a person who is just above a standard. This is the minimally qualified person who is above the standard, but only by a small distance on the construct of interest. The description of the characteristics of the minimally qualified person is represented symbolically as $\mathbf{minD_C}$. That is, $\mathbf{minD_C}$ is a refinement of $\mathbf{D_C}$ to focus on the subset of $\mathbf{D_C}$ that applies to the person just above the standard. $\mathbf{D_C}$ describes the characteristics of the range of persons above the standard. In a recent study on performance level descriptions, Egan, Schneider, and Ferrara (2012) used the labeled "Target PLD" for the $\mathbf{minD_C}$ with the following definition: "They [Target PLDs] define the expected performance of a student who just enters a performance level, and they represent the lower end of the Range PLDs." Further "they provide standard setting panelists with a common understanding of the sponsoring agency's expectations regarding the content-based competencies that students should possess in order to enter each performance level" (p. 93). Egan, Schneider, and Ferrara (2012) are considering the educational context for standards, but the same concept is used when setting standards for licensure and certification. There the description is of the minimally qualified candidate.

While many standard setting processes use the concept of $\mathbf{minD_C}$, there are many variations in the way that it is developed. Egan, Schneider, and Ferrara (2012) discuss developing the description of the minimally qualified person at the same time the $\mathbf{D}_C$ is developed, but then allow refinement during the standard setting process. In other cases, panelists produce the $\mathbf{minD}_C$ after they receive information about the $\mathbf{D}_P$ and $\mathbf{D}_C$ for the standard. In either case, the goal is to create a common understanding of the agency's intended standard so that it can be translated to the reporting score scale using the kernel selected for the standard setting process. If the panelists achieve a common understanding of $\mathbf{minD}_C$, they will be working to translate the same intended standard to the reporting score scale, and variations in the results will represent nuances in interpretation, as with any other type of translation between two languages.

As with the other definitions, it is useful to produce $\mathbf{minD}_C$ before the implementation of the standard setting method so that the agency can determine if it accurately reflects their intentions. Serious problems result when the standard setting process is near completion and the agency reviews $\mathbf{minD}_C$ and decides that it does not represent their intentions. Often the agency will want to adjust the value of the standard on the reporting score scale to make it closer to what they intended, but then, if they make an adjustment, the definitions that drive the process are not consistent with the final approved standard. A result is that it is difficult to produce a convincing validity argument for the level of the standard. Pellegrino, Jones, and Mitchell (1999) discuss such an adjustment for the standards set on the NAEP Natural Science Assessment.

As defined in Equation 2.6 in Chapter 2, $\mathbf{minD}_C$ = minimum ($\mathbf{D}_C \mid Ag$, $\mathbf{S}$, $\mathbf{D}_P$). That is, the definition of minimal qualifications to exceed the standard is a function of the standard intended by the agency, $\mathbf{S}$. Of course, the ideal case is never achieved in practice and distinct groups of experts will likely produce somewhat different versions of $\mathbf{minD}_C$. The differences in versions of $\mathbf{minD}_C$ are one source of variation in the final estimate of the standard on the reporting score scale. However, if $\mathbf{minD}_C$ is carefully developed to match the intentions of the agency, a qualified panel is selected to perform the translation, and an appropriate kernel is selected to help the panel with their task, then the estimated standard, $\hat{\zeta}_{Ag}$, should be consistent with the conceptual standard, $\mathbf{S}$.

3.5 Summary

This chapter has focused on the definitions that are critical to the estimation of a standard on a reporting score scale that is both consistent with the intentions of the agency that calls for a standard and the development of a convincing validity argument that supports the use of the standard. The definitions that are the basis of the standard setting process may vary in quality and precision of language depending on the characteristics of the agency that calls for the standard. Some agencies are very much involved with the standard setting process and approve every aspect of the process. The agency may also have the expertise to produce precise definitions. Other agencies focus on policy and may choose to delegate much of the standard setting process to others. An agency may not have the necessary expertise to produce detailed definitions that capture their intentions.

The theory presented in this book highlights the importance of clear policy, precise definitions, and careful review of all aspects of the standard setting process to make sure that all components are consistent with the intentions of the agency that calls for the standard. Neglecting these components of the standard setting process will likely result in estimated standards that are difficult to replicate, and the agency may view them as inconsistent with their intentions. Vague definitions also make it difficult to produce a strong validity argument to support the use of the estimated standard.

The process of standard setting has made great strides over the past few years. There is now a much greater understanding of the workings of the process. One area that has not received much attention is the influence of the characteristics of the agency on the results of the standard setting process. The agency is a difficult topic of study because they differ on many important features, and they may not want to be studied. However, it will be helpful if guidelines can be developed for agencies to help them play their very important role in the standard setting process.

4

The Construct and the Test Designed to Measure the Construct

4.1 The Target for Test Design and Development

The theory of standard setting presented in the previous chapters indicates that the technical procedure typically labeled "standard setting" is the translation of the intended standard (conceptually, as **S** and on a test construct as ζ_{Ag}) described in a policy definition, $\mathbf{D_P}$, to a point on the reported score scale on a test, $\hat{\zeta}_{Ag}$. The test score scale is the operational definition of the continuum implied by $\mathbf{D_P}$. In the technical literature on test design and construction, this continuum is often called a "hypothetical construct" because it is based on the idea that a hypothetical continuum exists, and that the continuum is a construction in the mind of the individual designing the test. In this book, the continuum is represented by the Greek symbol zeta, ζ, with the location on the continuum for individual *i* represented by ζ_i. When an item response theory (IRT) model is used to define the continuum, θ is used instead of ζ to indicate that the continuum has IRT scale properties, and τ is used when a true score theory model is used to define the continuum.

Although $\mathbf{D_P}$ is the main focus for the design or selection of the test used to operationalize the continuum assumed by the agency, Pitoniak and Cizek (2016) and Cizek and Bunch (2007) recommend that the content-specific definition, $\mathbf{D_C}$, be developed before a test is designed to measure the construct, or as a part of the test design process. Producing $\mathbf{D_C}$ early in the process should help ensure that inferences drawn from the test results will be consistent with $\mathbf{D_C}$ and the minimum qualifications to meet the requirements of $\mathbf{D_C}$, $\mathbf{minD_C}$. Consideration of $\mathbf{D_C}$ is particularly important when an existing test is selected for use in the standard setting process. It is possible, or even likely, that there will be skills and abilities described in $\mathbf{D_C}$ that are not required by the kinds of tasks included in an existing test, or a test selected rather than being designed, as the focus of standard setting. For example, many states are now using college entrance examinations as part of their state testing programs and need to set proficiency standards on these tests. But college admission tests were not designed to measure the

DOI: 10.1201/9780429156410-4

continuum implied by the $\mathbf{D}_\mathbf{P}$ or the specific content included in $\mathbf{D}_\mathbf{C}$. If that is the case, then something will need to be done to resolve the inconsistencies. Either $\mathbf{D}_\mathbf{C}$ and $\mathbf{minD}_\mathbf{C}$ should be adjusted to be consistent with the construct measured by the test, or a different test should be selected that is a better match to the requirements of the standard, or the selected test should be augmented to make sure all necessary inferences can be supported. The agency should approve the option chosen to make sure that the translation of the intended standard to the reporting score scale can be accomplished with an acceptable level of accuracy.

The test used with a standard setting process was likely designed and developed based on one of two different test design models. One design model is constructing a test to find the location of an individual on a well-defined unidimensional continuum. The second design model is to estimate the proportion of a well-defined domain of content that has been achieved by an examinee. In the ideal case, the design model used to develop the test should match the requirements of the construct implied by the $\mathbf{D}_\mathbf{P}$. Reckase (2017) provides a more complete discussion of these two models for test design.

While there are two basic models for test design, it is seldom that a test design follows each of these models in pure form. For the unidimensional continuum model, the goal of the test design process is to estimate the location of an examinee on a unidimensional continuum. Test tasks are designed to gain information from the examinee that can be used to efficiently determine the location of the examinee on that continuum. The domain model assumes that the test score represents the amount of a broad domain of skills and knowledge, and the goal of the test design process is to estimate the amount of the domain acquired by an individual. The estimate is based on performance on a sample of test tasks that represent the domain. The distinction between the two test design processes becomes less distinct when the test developer assumes a unidimensional continuum is a reasonable representation for the amount of skills and knowledge acquired from the domain, or when the developer decides that several types of test tasks are needed to accurately estimate the location on the unidimensional continuum. The tests selected to represent the continuum implied by $\mathbf{D}_\mathbf{P}$ are often based on a merger of the two development approaches.

The term "test task" has been used here to indicate the wide range of procedures employed to collect information from examinees that can be used to estimate their locations on the continuum of interest. Some examples of test tasks are multiple-choice test items, open-ended test items, essays, and computer simulations. Sometimes the more common term "test item" will be used when the test is composed of short test tasks.

Although the theory of standard setting presented here assumes that there is at least one intended standard that is located on the continuum that is implied by the $\mathbf{D}_\mathbf{P}$, neither the standard, or standards, nor the continuum is directly observable. Instead, a test is either selected or developed and the score scale for the test is the operational definition of the continuum implied by the $\mathbf{D_P}$. The standard setting process is used to translate the intended standard(s), as described in the $\mathbf{D_P}$, to a point, $\hat{\zeta}_{Ag}$, on the reporting score scale for the test. Use of the test and the estimated standard to make accurate decisions about individuals as meeting or not meeting the requirements of the $\mathbf{D_P}$ necessitates that a sound validity argument support the desired use of the test scores. Part of that validity argument should show that the test score scale is a good representation of the hypothetical construct implied by the $\mathbf{D_P}$. It is also important to note that the test can support other components of the standard setting process. For example, if more than one standard is to be translated to the reporting score scale, the test must define a scale that is broad enough to include the locations of all levels of the standards. The interaction of the test design and the standard setting process will be described in more detail when possible kernels for the process are described in Chapter 5. The remainder of this chapter will present information about the requirements for the test used as the operational definition of the continuum implied by the $\mathbf{D_P}$. The requirements for the test will be discussed within the context of the two types of test development models.

4.2 The Unidimensional Model for Test Development

Most standards on physical measurements are based on the unidimensional model for the construct of interest. For example, the Center for Disease Control and Prevention (CDC) has set a value of 30 or above on the Body Mass Index (BMI) scale as the standard for obesity for adults in the United States. There is an assumed hypothetical construct called "body fatness" and the location on the body fatness construct is estimated by the BMI (visit https://www.cdc.gov/obesity/adult/defining.html). However, persons with the same BMI can have different amounts of fat on their bodies. The BMI is essentially a single item test used to estimate the location of a person on the construct. More accurate estimates of location on the continuum can be obtained by combining the information from additional techniques to estimate the percentage of fat in the body makeup. From the perspective of a theory of standard setting, the issue is whether the use of the operational measures can be supported by a rigorous validity argument that differences on the reported scores represent differences in the continuum implied by the $\mathbf{D_P}$.

Standards for selecting employees, or selection of trainees for employment, use tests based on the unidimensional approach to test design and development more than do standards related to the outcomes of an educational program. For example, the Tennessee Valley Authority, a corporation that provides electricity to companies and community power utilities in the central United States, uses a battery of tests to select candidates for operations and maintenance training for their facilities (https://www.tva.com/Careers/Selection-Testing-for-Operations-and-Maintenance-Training). One of the tests in the battery measures spatial ability, the capacity to understand and use the relationships between shapes and objects. This ability has been found to be an important predictor of success in training courses for jobs such as pipefitter, welder, steelworker, and mechanic. These jobs require a person to visualize how parts fit together and how the parts are oriented in space.

One test that is used to operationally define the hypothetical construct of spatial ability is the Assembling Objects Test developed by the Edison Electric Institute (2001). This test consists of 20 items that require examinees to indicate whether shapes shown in drawings would fit together to make a larger shape. A second item type asks examinees to indicate which drawing properly connects two shapes with a line. The test tasks are not actual job activities, but they are indicators of a person's capabilities to learn the job tasks. There is no clearly defined domain of such test tasks, so there is no process for sampling the tasks from the domain. Instead, test items were written and pre-tested on the target population of examinees to determine if they can accurately provide information about the location of persons on the hypothetical construct. The critical evaluations of the test results are whether the items are homogeneous (matching the unidimensional model), and whether they provide accurate estimates of persons' locations along the reporting score scale. The evaluation of accuracy is typically done using studies of test reliability for the population of interest. With the widespread use of IRT, accuracy may also be shown through estimation of conditional standard errors of measurement or amount of information provided at points along the score scale.

When the score scale for a test is used as the operational definition of the hypothetical construct implied by the $\mathbf{D}_P$, it needs to meet the usual standards of quality as described in the *Standards for Educational and Psychological Testing* (AERA, APA, NCME, 2014). That is, there must be evidence that inferences drawn from the test results can be supported and that the reported scores are accurate reflections of the intended construct. In the context of standard setting, there are some additional requirements. First, it must be possible to do the tasks specified by the kernel selected for the standard setting process with the types and number of test tasks used in the test design. For example, some standard setting kernels are designed for use with multiple-choice test items and will not work well with essay questions. This implies that the kernel should be selected to match the test used to operationally define the continuum implied by $\mathbf{D}_P$, or the test should be designed to match the kernel that has been selected for use in the standard setting process.

Second, the test must be designed so that the scores are sufficiently accurate in the range of the score scale where the estimated standard is likely to be. That is, if the standard is likely to be for minimal competence, the test scores should be reasonably accurate for low-performing examinees. If the $\mathbf{D_P}$ indicates that the intended standard is rigorous, the test should be designed to give accurate estimates at the upper end of the scale. Further, the test should be designed to provide high decision consistency once the location of the standard has been estimated. This is especially important if the $\mathbf{D_P}$ indicates that multiple standards are to be translated to the reporting score scale. In that case, the test should be designed to give accurate estimates of the location of the examinees in all the regions defined by the standards.

An important concept in the current technology for test design is that test tasks (test items) provide information about examinees' capabilities for a range of the hypothetical construct underlying the test, but not for all parts of the continuum underlying the test score. The test tasks focus on a part of the continuum. This is especially true for test tasks with two score categories such as 1 for a correct response and 0 for an incorrect response. The concept of information in IRT is based on how well test tasks can distinguish between individuals located near each other on the continuum underlying the test (see Lord (1980) for derivations of item and test information). Thus, to accurately determine whether a person is above or below a standard, a test should have enough test tasks focused on the range of the continuum that contains the location of the standard so that there will be sufficient information for accurate estimation of the location of examinees who are near the standard.

For tests designed to measure a single continuum, like the spatial relations test described above, the original approach to test design was to find test tasks that had a high correlation with a performance measure, such as performance in a training course, and somewhat lower correlations with each other. This approach was designed to optimize the decision accuracy of the test for selecting persons for the training course. Nunnally and Bernstein (1994) provide step-by-step guidelines for producing tests in this way (see pages 327–329). Their goal was to see that the test was a reliable measure of the construct of interest and focused on the range of the continuum important for the decision.

With the widespread use of IRT, the selection of test tasks has changed, but with the same goal as when traditional methods had been used. First, test tasks are administered to a representative sample of examinees from the target population for the purpose of collecting item response data that can be used to determine the characteristics of the test items. These data are used to calibrate the items with the IRT model that is selected for use. The result of the calibration will give information about the qualities of the test items (difficulty, discrimination, etc.) and the fit of the IRT model to the data from the item responses. Then, items that are well fit by the IRT model are selected for further consideration. The calibration process also defines the continuum underlying the test. When using a test as the basis for the standard setting process, it is important that it provide good accuracy of estimation for person locations in the range of the continuum where decisions will likely be made. This can be assured by selecting items so that they provide sufficient

IRT-based information in that range. Because the IRT concept of information is inversely related to the standard error of estimates of examinee location, this approach also ensures that there will be good classification accuracy in that range of the scale. Luecht (2006) provides step-by-step procedures for designing tests to give good decision accuracy at the estimated location of a standard on the IRT-based reporting score scale.

The important outcome of both the classical test theory and IRT-based test design and development processes is that test tasks are produced that are targeted at the location of the standard. Such targeting will support both the accurate translation of the standards implied by the $\mathbf{D_P}$ to the reporting score scale and the accurate classification of individuals as above or below the estimated standard. Further, there will be evidence supporting a convincing validity argument for the inferences made using the standard.

4.3 Domain Sampling Model for Test Development

In many cases, standards are set to define the desired amount of skills and knowledge that a person should acquire through an educational or training program. There is a domain of skills and knowledge that is the target of instruction or training, and the $\mathbf{D_P}$ indicates the minimum of that domain a person should have acquired to have properly profited from the instruction or training. The standards set on the achievement tests used by states to judge the quality of student learning are typically based on this premise. There is a state-mandated or state-recommended curriculum for each grade level, and standards are set to indicate when students have acquired enough knowledge and skills to move on to the next grade level of instruction.

The same approach to test design and development used for school achievement tests is often used for licensure or certification tests, but the domain of knowledge and skills is often determined through a job analysis for the jobs that are likely for a newly certified person. There is an expectation that training programs will develop those skills and knowledge, and the $\mathbf{D_P}$ specifies how much is enough to have a reasonable likelihood of success in the entry-level position.

Consistent with the theory presented in this book, this approach to test design and development begins with the continuum implied in the $\mathbf{D_P}$. For example, the $\mathbf{D_P}$ for an eighth-grade mathematics test might indicate that students in eighth grade should learn enough of the state-mandated mathematics curriculum for that grade to understand the concepts presented in the algebra course at grade 9. Another example might be that the $\mathbf{D_P}$ is about the knowledge and skills needed for an entry-level automotive mechanic who will work on exhaust systems. The definition implies that there is a domain of knowledge and skills about exhaust systems that is appropriate for the types of automobiles that will be serviced by an entry-level person. It is expected that the person wanting certification for work on exhaust systems has gained the necessary knowledge and skills either through a training program or experience working with a qualified person.

The theoretical basis for this approach comes from the area of statistical sampling theory. If a random sample is drawn from a population, then the mean of that sample is an unbiased estimate of the mean of the population. From the perspective of test design, the population is the set of all test tasks that could be developed to assess a person's knowledge and skill of the domain. If a random sample of those test tasks is drawn and administered to a person, and the proportion of correct responses is computed, that proportion is an estimate of the proportion of the entire domain that the person can answer correctly. This straightforward statistical concept is difficult to apply in practice because the full population of relevant test tasks does not exist, so it is not possible to randomly sample from it. Instead, test developers partition the domain into parts and then produce test items to assess the parts. The goal is to get a representative sample of the domain even though it is not a strict random sample. Webb (2006) describes the process this way.

> In general, the content to be included in a test needs to be clearly identified by defining decision rules for the development and selection of test items. Such decision rules can be used to define the domain, or universe, of all possible items measuring students' content knowledge related to the learning expectations and to identify a procedure for selecting tasks so that the desired inferences can be made.
>
> **(p. 163)**

When the procedure described by Webb (2006) is followed rigorously, a specified number of items need to be developed and included in a test for each content category or topic. For example, content for a mathematics test might include (1) geometry and spatial sense, and (2) algebra and functions so test items would need to be included for each of these content areas. This approach leads to a test design that is based on multiple knowledge and skill dimensions rather than the unidimensional model discussed in the previous section. It is certainly possible for students to learn algebra and functions without learning geometry and vice versa. Yet, $\mathbf{D_P}$ may indicate that the standard is on the proportion of the domain, indicating that the standard is set on a single continuum rather than on multiple dimensions.

Webb (2006) further states

> If a state test is designed to determine whether students have achieved at the level identified in state standards, and if state law requires it, the depth-of-knowledge level of each of the test items should be the same as the depth-of-knowledge level of the standard or objective the test item is designed to measure. (p. 166)

Following such a procedure for test design and development indicates that the reported score on the test is not representing the proportion of the full domain that someone knows, but instead represents the proportion of the domain that is at a level of challenge that matches the $\mathbf{D_P}$ for the standard. This has implications for the level implied by the translation of $\mathbf{D_P}$ to the reporting score scale.

Consider the following case. Suppose that the full domain defined by the content of instruction is relatively easy and the average examinee can answer about 80% of the items from the full domain. But suppose that the $\mathbf{D_P}$ specifies a very high standard. Then, according to Webb (2006), the test should be composed of the difficult items from the domain so the average examinee can answer a much smaller percent of them correctly. The persons doing translation from the $\mathbf{D_P}$ to the point on the reporting score scale will need to understand the differences in a representative domain sample and the sample taken from a restricted range of the domain.

When multiple standards are indicated by the $\mathbf{D_P}$, a test design that focuses on a single level of difficulty (depth of knowledge) might not have accurate indications of performance in the range for some of the standards. Webb (2006) acknowledges this possibility.

> Most standards-based assessments are designed to identify the level that students have attained vis a vis a standard, or standards, using categories such as *below basic, basic, proficient,* and *advanced*. ... Because of this, a test designed to more accurately identify the achievement of individual students needs to provide some range of the complexity of the items. Without definition of such a range, the information about students produced by the test is of limited value.
>
> **(p. 168)**

Test designs for the measurement of the proficiency of examinees on a domain of skills and knowledge have great variations. The variations depend on how the persons constructing the test relate the requirements for the test to their interpretation of the details of the $\mathbf{D_P}$. But the $\mathbf{D_P}$ might not be considered at all. Sometimes the tests are designed without regard to the fact that the reporting score scale will be used as the operational definition of the continuum implied in the $\mathbf{D_P}$. The agency might select an existing test for use or set policy that the test design should focus on purposes unrelated to comparing examinees to the standard.

4.4 Implications of Test Design for Standard Setting

The information in the previous sections indicate that development of a reporting score scale can be done in many different ways. Because standard setting is conceptualized as the translation of $\mathbf{D_P}$ to that reporting score scale, it is critical that the persons doing that translation have a clear understanding of how the reporting score scale was developed. This implies the need for training, especially if the persons involved do not already have intimate knowledge of the test development and scoring process. Without that training, the standard setting process is like translating between two languages when the persons doing the translation have native proficiency in one language and a tourist level in the other. A humorous AT&T Wireless television commercial (https://www.ispot.tv/ad/o6VN/at-and-t-wireless-ok-translator) titled "OK: Translator" provides a good example. A translator translates his client saying, "We need this merger" with another company, as "My client needs a hug." When this happens in standard setting, the translation to the reporting score scale will be poor and certainly not capture the nuances of the $\mathbf{D_P}$.

Another issue of test design and selection is the amount of information the test provides about examinees in the region of the reporting score scale near the intended standard. Kane (1994) suggests that the test should be designed with the possible location of the standard in mind so that there will be high precision in the region around the cut score. If the intended standard is high on the continuum implied by the $\mathbf{D_P}$, then test must have enough items that can differentiate levels of performance in that region of the continuum. From a test design perspective, there must be a number of relatively difficult items to define differences on the reporting score scale at that level. If the test lacks those items, it will not be possible to accurately translate $\mathbf{D_P}$ to the reporting score scale. It would be the equivalent of translating a literary document from one language to another when the translator has only an elementary school level of language proficiency for the second language. The vocabulary and sentence structure knowledge will not be available to accurately capture the meaning of the complex, nuanced document.

The same issues apply to cases when the intended standard is low on the continuum. The test must have enough relatively easy items to result in a reporting score scale that can support accurate translations of $\mathbf{D_P}$ in that region of the scale. This is a particularly critical issue when the test is composed of multiple-choice test items. A poor translation of $\mathbf{D_P}$ could result in a point on the reporting score scale that is lower than the score resulting from randomly guessing on every test item. This would be a meaningless translation that could not support any useful inference about the location of persons relative to the intended standard.

4.5 Summary

The test that is used with a standard setting process is a critical component. It defines the "language" that is the focus of the translation from the $\mathbf{D_P}$.

There needs to be a solid validity argument that supports inferences about examinees' locations on the continuum implied by $\mathbf{D}_P$ from their performance on the test as indicated by their locations on the reporting score scale. Of course, such inferences require that the test provide reliable measurement of the population of interest. It also requires that the level of precision be at a sufficient level in the region of the reporting score scale that corresponds to the location of the intended standard on the hypothetical construct for the continuum implied by $\mathbf{D}_P$.

There are many different ways to design a test, and the design defines the language of the reporting score scale that is the focus of the standard setting process. It is important that the persons involved in the standard setting process understand the language of the reporting score scale so they can do a proper translation. This means that training on the meaning of the reporting score scale is an important part of the standard setting process, especially if the persons involved are not intimately involved in the use of the test. They need a nuanced understanding of the performance indicated by points along the scale. This will require information about the test design and development process. It is common practice for standard setting processes to have panelists take the test and score the results so they will have a good understanding of the skills and knowledge required to perform on the test. Panelists may also be led through a detailed discussion of the conceptual model underlying the development of the test (the domain definition or a description of the unidimensional construct) and the connection of the development model to the content-specific definition, $\mathbf{D}_C$.

Ignoring the details of the test design and the development and/or training of participants about the meaning of the reporting score scale will likely result in a poor translation of $\mathbf{D}_P$ to $\hat{\zeta}_{Ag}$. It would be like asking persons to translate an important document from a language they know well to a language of which they have only minimal knowledge. The result is likely to be so inaccurate as to be unusable. The most serious outcome is that the user of the result might not know that it is a poor translation and accept it as accurate. Good test design and development, and good training, will insure against poor translation. Of course, the other parts of the standard setting process must be of quality as well.

5

The Kernel of the Standard Setting Process

5.1 Introduction

The literature on standard setting considers the kernel of the standard setting process as "*the* standard setting process." For example, the classic book edited by Cizek (2012) has a section titled "Standard Setting Methods," which is a compendium of what are called "kernels" in this book—modified Angoff, body of work, bookmark, etc. Here, however, the kernel is considered as one component of a complex standard setting process. It is a key component because it generates the data that are used to estimate the point on the reporting score scale that corresponds to the intended standard implied by the $\mathbf{D}_P$. But it is difficult to argue that it is more important than developing a precise $\mathbf{D}_P$, good training procedures, or selecting qualified panelists.

In this book, the kernel of the standard setting process is defined as the specific task that panelists perform to provide information that can be used to estimate the location of the standard on the reporting score scale for the test. The kernel is represented by the symbol *K*. The kernel can be thought of as a function of a panelist, the Test Specifications, the policy definition, $\mathbf{D}_P$, the content-specific definition, $\mathbf{D}_C$, and the definition of the minimally qualified examinee, $\mathbf{minD}_C$. The outcome of the application of the kernel function, $\mathbf{R}$, is information used to estimate the location of the intended standard, $\hat{\zeta}_{Ag}$, on the reporting score scale:

$$\mathbf{R} = K\left(\text{Panelist, Test Specifications}, \mathbf{minD}_C, \mathbf{D}_C, \mathbf{D}_P\right). \quad (5.1)$$

$\mathbf{R}$ is a general symbol for the information provided by a panelist when performing the tasks specified by the kernel. Some standard setting procedures may use more than one kernel, so a subscript may be included to indicate which kernel is being considered, K_1, K_2, etc. For example, panelists may be asked to perform bookmark placements for two rounds of the standard setting process and then directly estimate a location on the reporting score scale.

To help distinguish the kernel from the rest of the standard setting process and to emphasize its function in the translation of policy to a point on the

DOI: 10.1201/9780429156410-5

reporting score scale, a fanciful example of language translation is described, followed by a parallel example of a standard setting process. Suppose there is a need to translate a paragraph from English to Spanish, but no professional translators are available to perform the translation, so an alternative approach has been developed to get the translation. Ten individuals are recruited who are proficient in English and who know some Spanish. Some of the individuals learned Spanish in secondary school years ago, while others learned some Spanish for the purpose of travel to a Spanish-speaking country (i.e., tourist Spanish). These individuals are divided into two groups of five to work together on the project.

The individuals are brought together in a room to learn about their task and the steps they need to follow. They are told how their translations will be used and the background of persons who will be reading their translations. Then they are divided into the two groups and are asked to do a word-for-word translation of the paragraph into Spanish. If they do not know the corresponding word in Spanish, they can leave that result blank. After they complete this task, the results are collected and compiled into a form that shows the Spanish words selected for each English word for the five persons in the group. This information is provided back to the group along with English/Spanish dictionaries for their use. They are instructed to review the results and discuss the word selections and each person can decide for themselves if they want to change the word they selected as their translation. The translations are then collected from each person.

Because word-for-word translations often generate awkward sentences in the target language, a person with native proficiency in Spanish, but who is not proficient in English, reads the sentences in Spanish and edits them for grammar and sentence structure. The edited sentences are then returned to their authors for review, and then they are discussed within the five persons in the respective groups. They are instructed to decide if one sentence is preferable over another and if they want to make changes to the sentences. The results of their deliberations are submitted as their best attempt at the translation. The results from the two groups can be compared to determine if they generated comparable translations of the paragraphs. In theory, the results should be similar, but it is unlikely that they will be identical.

Because this is a fanciful example, an additional piece of information is included. Suppose that at the end of this process the person who initiated the process discovers that the paragraph in question has already been translated into Spanish by a professional bilingual translator who works for the United Nations and is well known to produce accurate, highly nuanced translations. If the translation process with the ten individuals was a good one, the results should be a good match to the result from the professional translator. If the match is poor, the group translation process is suspect.

In this fanciful example, the word-for-word translation of the paragraph is the kernel of the process. That is the main task done by the participants and their translated words are the outcome of the process, **R**. The editing of the sentences is the equivalent of the process used to convert **R** into the standard. In the standard setting process, the equivalent to the edited sentence is the estimated point on the reporting score scale. The editing process is the equivalent of the estimation procedure used to identify the estimated standard for a panelist. In the fanciful example, the result of a professional translation is also available. That is the equivalent of the intended standard from the agency that produced the paragraph.

The fanciful example is a good match to many standard setting kernels that collect data using an item-by-item rating procedure such as various Angoff-type methods. For those methods panelists rate each item (like translating each word), and often get feedback about the observed difficulty of test items (like getting the English/Spanish dictionary). However, it is not a good match to other types of kernels. It is left to the reader to imagine how they could change the example to match other types of kernels.

The example shows that the kernel is a means to collect information that can be used to obtain the final estimate of the translation. There is also an intended translation, but, unlike the example, the intended translation is never precisely known. All that might be known is that the agency who called for the standard approved it as matching their intentions or judged it to be inconsistent with their intentions and either rejected the result outright or chose to modify it to better match their intentions.

The remainder of this chapter provides examples of some commonly used kernels. This is not meant to be a comprehensive listing of currently available kernels. There are other books available that focus on describing standard setting kernels (see Cizek (2012) as one example). Also, every standard setting kernel is modified in some way when it is used in practice to meet the requirements and limitations of a particular situation. Therefore, it is important to determine the specific tasks that panelists are asked to do rather than expect that all kernels that have been given the same descriptive label are exactly the same. It is also the responsibility of those reporting on a standard setting process, either in a research article or the technical report for a process, to provide the details of the process and the tasks required of panelists so that others can accurately understand the steps in the process.

5.2 Examples of Standard Setting Kernels

The examples that are presented here make several assumptions about the full standard setting process so that the focus can be on the details of the kernel. It is assumed that there is a detailed policy definition, $\mathbf{D}_P$, and that a test has been produced or selected that does a reasonable job of locating examinees on the construct implied by $\mathbf{D}_P$. That is, there is a credible validity argument for the use of the test with the $\mathbf{D}_P$. It is also assumed that the

$\mathbf{D_P}$ has been accurately translated into the content definition, $\mathbf{D_C}$, which is consistent with the specifications for the test. Finally, it is also assumed that $\mathbf{minD_C}$ accurately describes the minimally qualified examinee in a way that is consistent with $\mathbf{D_P}$ and $\mathbf{D_C}$.

Given all the initial components of the standard setting process, it is assumed that the panelists have been given training on all the definitions and have reached a common understanding of those definitions. The panelists have also taken the test, scored it, and understand the test specifications for the test. The panelists are also familiar with the examinee population.

The fact that there are so many assumptions before discussing the use of the kernel emphasizes the complexity of the standard setting process and the relatively modest part that the kernel plays in the full process. Consumers of information from standard setting processes, and those who design and implement the processes, should focus on all aspects of the process rather than emphasizing the impact of the kernel. The kernel is important, but other aspects are equally important.

5.2.1 The Bookmark Kernel

The Bookmark kernel is widely used for translating policy to points on the reporting score scale for many education applications. This is likely because the method is considered as easy to understand and it does not take extended time to implement (Cizek & Bunch, 2007). As is true of all the standard setting kernels, there are many variations in the application of the Bookmark kernel. The initial version of the Bookmark kernel (Lewis, Mitzel & Green, 1996) will be presented first and then some of the variations will be described. Some of the issues that can reduce the accuracy of the estimates of points on the reporting score scale when the Bookmark kernel is used will also be discussed in this chapter.

The theoretical bases of the Bookmark kernel are the same as those for a very general conception of item response theory (IRT) (Lord, 1980). That is, it is assumed that there is a hypothetical continuum that is the target of measurement for the test and that the goal of the test is to estimate the locations of individuals on that continuum. It is also assumed that the interaction of test items and persons can be represented by a mathematical function that gives the probability of obtaining a selected item score for persons at various locations on the underlying continuum. In most cases, a convenient mathematical equation is selected to represent the interaction of persons and test items. Lewis, Green, Mitzel, Baum, and Patz (1998) indicate that they typically used the three-parameter logistic (3PL) model to represent the interaction for multiple-choice test items and the two-parameter partial credit model (PCM) for test items with multiple score points. However, the conceptual model for the Bookmark kernel can be implemented with any mathematical model that is a good representation of the interaction between the examinees and the test items.

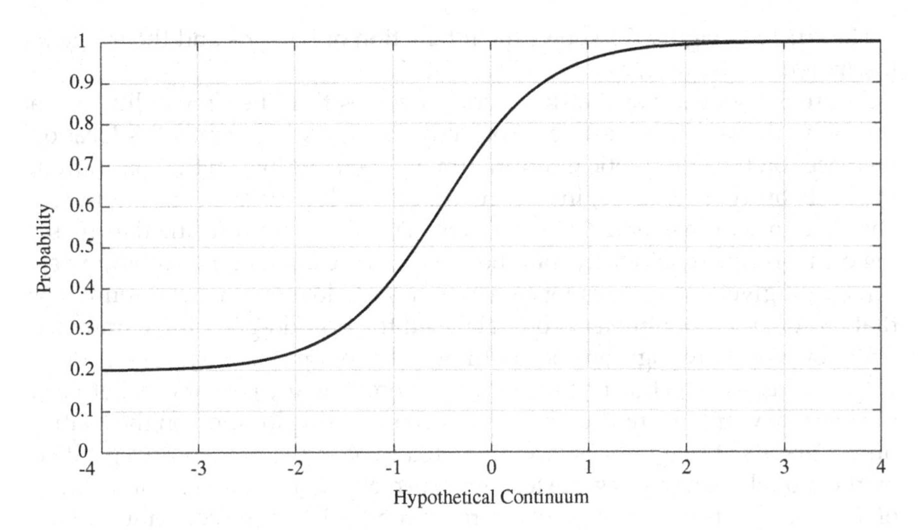

FIGURE 5.1
Item characteristic curve (ICC) for the three-parameter logistic model with parameters $a = 1.1$, $b = -0.5$, and $c = .2$.

The mathematical representation of the interaction between examinees and a test item is shown in Figure 5.1 for an item accurately modeled by the 3PL model. The curved line shown in Figure 5.1 is called an item characteristic curve (ICC) because it shows the way a test item functions for examinees located at points along the hypothetical scale. The ICC in Figure 5.1 represents an item of moderate difficulty. The mathematical formula for the 3PL model is given by:

$$P_j(\theta) = c_j + (1 - c_j)\frac{e^{1.7a_j(\theta - b_j)}}{1 + e^{1.7a_j(\theta - b_j)}}, \tag{5.2}$$

where $P_j(\theta)$ is the probability of a correct response to test Item j for a person located at θ on the hypothetical continuum,

c_j is the lower-asymptote parameter that is usually interpreted as the likelihood that the person can guess the correct response to the test item,

a_j is a parameter that indicates how quickly the probability of correct response changes with change in value of θ—often called the slope or discrimination parameter of the test item,

b_j is a parameter that indicates the location of greatest rate of change of the probability of a correct response with change in value of θ—often called the difficulty or location parameter for the test item,

and 1.7 is a constant value that makes the results from the model approximate those from the normal ogive model—a model of historic interest based on the normal distribution function.

The curve in Figure 5.1 shows the interaction of persons and the test item for an item with $c_j = .2$, $a_j = 1.1$, and $b_j = -0.5$.

Figure 5.1 shows that the IRT model specifies that the probability that a person will answer the test item correctly increases as the person's location increases on the hypothetical continuum. For persons low on the continuum, the probability is not zero, indicating that those low on the continuum still have a chance to answer the item correctly. This is usually attributed to guessing on a multiple-choice item, but there are many explanations for why a person might give a correct response when they are low on the continuum. For that reason, the *c* parameter is often labeled the pseudo-guessing parameter.

Persons who are high on the continuum have a probability very close to 1.0 of giving a correct response to the test item. For any probability between the lower asymptote and 1.0, there is a corresponding location on the continuum. Persons at that location are predicted to have the probability specified by the model of answering the test item correctly. For example, a probability of .7 corresponds to a location of approximately −0.2 on the continuum. That means that the IRT model predicts that persons who are located at −0.2 on the scale have a probability of .7 of answering the test item correctly. If there were 100 persons at −0.2 on the scale, 70 would be predicted to answer the test item correctly. Of course, this is a statistical prediction based on the IRT model and different samples of 100 persons at −0.2 on the scale would have slightly different numbers of persons giving a correct response. On average, over many samples of 100, the mean number of persons with correct responses should be very close to 70. These inferences about examinee performance assume that the model is a realistic representation of the interactions of the examinees and the test item.

This IRT conceptualization of the interaction of examinees and test items is critical to the theory behind the Bookmark kernel. The Bookmark kernel first requires the selection of a probability of correct response to the test item that is the boundary between examinees being considered successful on the item and being unsuccessful. In a sense, the selected probability is a definition of expected performance for examinees who have the skills and knowledge required to answer the test item correctly. Lewis, Mitzel, and Green (1996) suggested using a probability of 2/3 (i.e., .67) for this purpose because panelists understand the fraction and the value indicates that persons who have the skills and knowledge required by the test items would not have to be able to correctly answer the test item with certainty.

Once this probability is selected, the task for panelists when using the Bookmark kernel is to identify the item that has the selected probability of correct response for persons who match the description in $\mathbf{minD_C}$. To help the panelists do this task, Lewis, Mitzel, and Green (1996) organized the items on the test that operationally defines the continuum implied by the $\mathbf{D_P}$ into a booklet with one page per test item. They also arranged the pages in the booklet in the order of the location of the items on the continuum corresponding the selected probability, often called the *mapping probability*.

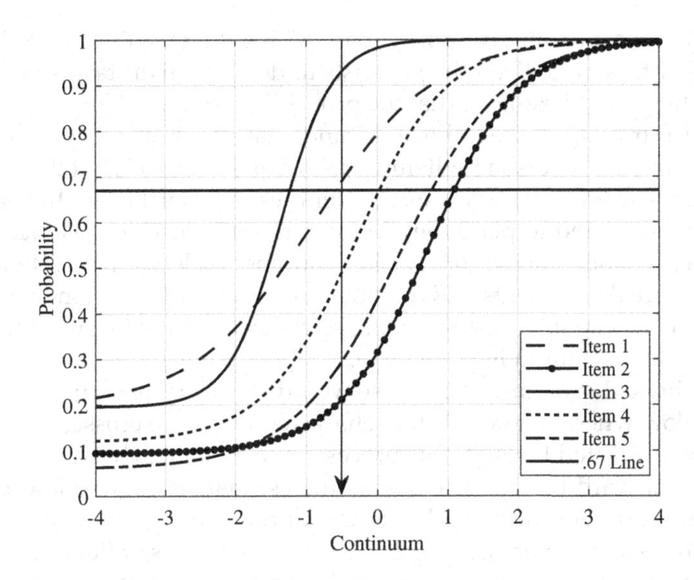

FIGURE 5.2
Item characteristic curves for a five-item test

Further, because they believed that panelists might have difficulty judging the influence of guessing on the probability of a correct response to the test items, they estimated the locations of the items after removing the hypothetical influence of guessing from the determination of the locations of the items. The booklet containing the ordered test items used with the Bookmark kernel is called the Ordered Item Booklet (OIB). A small example is given next to help make this process concrete.

Suppose that a five-item multiple-choice test has been well calibrated using the 3PL model. Figure 5.2 shows the ICCs for the test items. The figure indicates that Item 3 on the test is easiest because examinees on the continuum above –1 have over a .8 probability of answering this item correctly. Item 2 is the most difficult. Examinees need to be above 2 on the continuum before having a high probability of a correct response. Note that the item numbers are the order of the items as they were placed in the test booklet administered to examinees. The figure also shows a horizontal line at .67, the mapping probability for the items. The ordering of the items in the OIB for the Bookmark kernel would be according to where the items crossed the .67 line. That ordering would be Item 3, Item 1, Item 4, Item 5, and Item 2. The arrow in the graph represents the intended standard implied by $\mathbf{D_P}$. It is at –0.5 on the continuum. Of course, this value is not known by the panelists. It is the value estimated through the translation process from $\mathbf{D_P}$ to $\hat{\zeta}_{Ag}$, or in this case $\hat{\theta}_{Ag}$ because an IRT model is used to define the reporting score scale.

The task for the panelists when using the Bookmark kernel is to work through the OIB from the easiest to the hardest item and indicate the first item that has a probability of less than .67 for the examinee described by $\mathbf{minD_C}$. For this example, if the panelists understand their task and the definitions, they should estimate that the probability for Item 3 for a person with capabilities $\mathbf{minD_C}$ as well above .67 and that the probability of a correct response for such a person for Item 1 was slightly above .67, but the probability for Item 4 was clearly below .67. Under these circumstances, the panelists would be instructed to place their bookmark on Item 1, the last item in the OIB with probability above .67. The item numbers selected for the bookmark placement are the elements of **R** in Equation 5.1. The information in **R** is used to estimate a location on the reporting score scale for the test used to operationalize the continuum implied by $\mathbf{D_P}$.

The estimated standard in this case would be the value on the continuum that is below where the Item 1 item characteristic curve crosses the .67 line. The value estimated through this process is –0.59, somewhat lower than the intended standard of –0.5. The reason the estimate is slightly lower is that there is no test item in the OIB that has a probability of .67 at the value of –0.5 on the θ-scale. There is a gap between the locations of Items 1 and 4. Of course, if there were more items included in the OIB and if their locations were evenly spaced on the continuum, any gaps would be small, and the estimated standard would be very close to the intended standard.

The example as presented is not an exact match to the Bookmark kernel described in Lewis et al. (1998). In that study, the authors recommend adjusting the probabilities from the 3PL-model to remove the influence of the pseudo-guessing parameter, *c*. They adjust the probability of correct response to the test items using the following equation:

$$P_j^*(\theta) = \left(P_j(\theta) - c_j\right) / \left(1 - c_j\right). \tag{5.3}$$

Figure 5.3 shows the ICCs with this adjustment.

Figures 5.2 and 5.3 are different in a few ways. First, in Figure 5.3, all the ICCs have lower asymptotes of 0 instead of c_j, as in Figure 5.2. In this case, the ordering of the items based on the intersection with the .67 line is the same for the two graphs. The match of the intersections is not a mathematical certainty. In rare cases, the ordering of the test items can be different for the two representations of the ICCs. For this simple example, there is no change in ordering in the OIB when the influence of the lower asymptote is removed. The most crucial difference is that Item 1 now has less than a .67 probability of a correct response at the intended standard of –0.5, so a panelist who understands the process should place the bookmark on Item 3. This would result in an estimated standard of –1.11 on the θ-scale instead of the intended value of –0.5 as shown by the arrow. This is a substantial underestimate of the intended standard that is the result of the large gap between the curves for Items 3 and 1.

The results from this simple example show the basic process for the Bookmark kernel, and it highlights the importance of having sufficient items

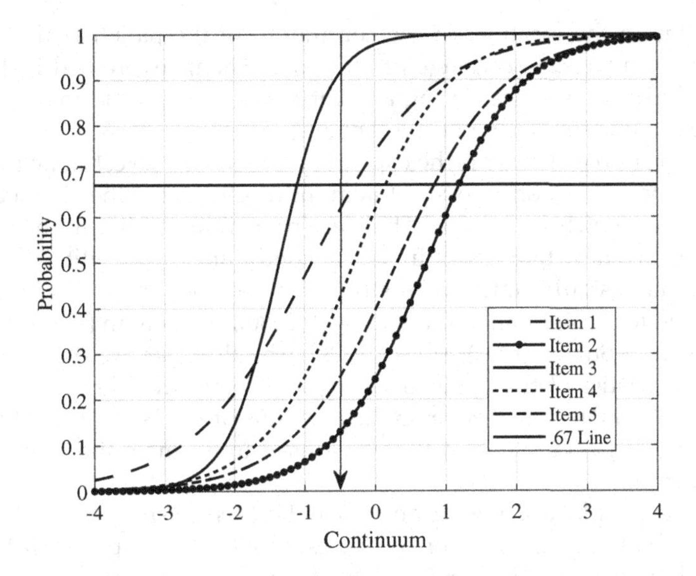

FIGURE 5.3
Item characteristic curves for a five-item test corrected for the magnitude of the lower-asymptote parameters.

in the OIB and that the items be evenly distributed along the continuum defined by the reporting score scale. Gaps in the locations of test items are particularly important because panelists cannot provide information that will yield an estimate of the standard in the region defined by a gap. There are solutions to the "gap" problem, but before discussing the solutions, it is important to consider how the Bookmark kernel functions under less ideal conditions, when there is error in the item parameter estimates, and when panelists are not perfect estimators of the probability of correct response to the test items.

In an operational standard setting process, there are numerous sources of error in translation from $\mathbf{D_P}$ to $\hat{\zeta}_{Ag}$. The presence of these sources of error is one of the reasons why standard setting processes use a number of panelists for the translation process. The hope is that by using multiple panelists and then averaging the results, the final estimate will be close to the standard intended by the agency as expressed in $\mathbf{D_P}$. The first source of error is variations in panelists' understandings of $\mathbf{D_P}$, $\mathbf{D_C}$, and $\mathbf{minD_C}$. In the simple example presented here, the intended standard was at –0.5 on the continuum, but different understandings of the definitions might lead panelists to believe that the intended standard is in some range around –0.5—some higher and some lower. This is a result of the imprecision of the language used to describe the desired skills and knowledge, and possibly due to insufficient training and discussion of those definitions to get a solid common understanding of the meaning of the definitions. Some researchers, such as Peabody and Wind (2019), suggest that the different interpretations might also be due to the location on the continuum of the panelists themselves. Panelists with weak knowledge of the requirements expressed in $\mathbf{D_C}$ may interpret $\mathbf{minD_C}$ as representing a point lower in the scale than panelists with greater understanding of the content.

A second source of error is the capability of panelists to estimate the probability of correct response to a test item in the OIB given the characteristics of examinees matching $\mathbf{minD_C}$. There are numerous studies in the research literature that show that the estimates of conditional probabilities of correct response have significant amounts of error and they may also be statistically biased (e.g., Brennan and Lockwood (1980), Tinkleman (1947), Truxillo, Donahue, and Sulzer (1996), Wyse (2018)). The third source of error is that the characteristics of the items are estimated rather than known. The item parameter estimates have error so the locations of the items in the OIB may not be perfectly accurate, and the value used to estimate the standard has some error in it.

The same simple, five-item example can be used to show the influence of these different sources of error on the estimation of the point on the continuum that corresponds to $\mathbf{minD_C}$. First, suppose there are two panelists: one that interprets $\mathbf{minD_C}$ as consistent with –.3 on the θ-scale and the other interpreting it as consistent with –.7. These two values were selected for the example because the average is –.5, the unobserved intended standard. The question is whether the average of the estimated standards for these two panelists yields –.5 when they properly use the Bookmark kernel.

From Figure 5.3, both panelists would estimate Item 3 as having a probability of correct response for their interpretation of $\mathbf{minD_C}$ of greater than .67. However, for the panelists interpreting $\mathbf{minD_C}$ as –0.7, the second item in the OIB is estimated to have a probability of correct response of less than .67, so they place the bookmark on the first item of the five and get an estimated standard of –1.11. The second panelist estimates the second item in the OIB to have a probability slightly above 2/3, but the third item has a probability well below 2/3, so they place the bookmark on the second item and get an estimated standard of –0.31. The average of these two estimates of the standard is –0.71. This is not equal to the intended standard of –0.5, but it is less of an underestimate than the estimate from the first panelist. The errors in interpretation of $\mathbf{minD_C}$ in opposite directions tended to compensate for the gap between the items, but it did not completely adjust for that problem. The example does show the value of using multiple panelists to compensate for different interpretations of $\mathbf{minD_C}$. Of course, both panelists could have had misinterpretations in the same direction. An important result of the example is that errors in interpretation of $\mathbf{minD_C}$ can result in differences in placement of the bookmark.

The second source of error when using the Bookmark kernel is the estimation of the probability of correct response to the items for persons located at the point on the scale that corresponds to $\mathbf{minD_C}$. No thorough studies have been found in the research literature on standard setting that indicate how well panelists can estimate the conditional probability of correct response, but there are many sources that state that panelists cannot estimate probabilities very well. The amount of error modeled for this simple example is that panelists have an error distribution with standard deviation .1 when estimating a true probability of .5. That means that their estimates range from roughly .2 to .8 (±3 standard errors), with most between .4 and .6. For probabilities that are very high or very low, the range is not symmetric because it is impossible to go three standard errors above .9 or below .1. For that reason, the error is modeled with a beta distribution. That distribution has two parameters, a and b. The a parameter is set at 12 and the b parameter is set at $b = 12(1-p)/p$. This gives the desired distribution for $p = .5$ and comparable distributions for more extreme values of p.

Table 5.1 gives the item numbers in the order used for the OIB. It also provides the conditional probabilities from the IRT model for the intended cut score of –0.5. These can be interpreted as the probability of a correct response for examinees at exactly –0.5 on the θ-scale for the IRT model. Columns three to seven provide the hypothetical estimated probabilities by five simulated panelists using the error model described in the previous paragraph assuming they accurately interpret $\mathbf{minD_C}$. All the simulated panelists estimate the first item in the OIB as having greater than a 2/3 probability of a correct response, so the inaccuracies of their estimates do not have the serious consequence for the Bookmark kernel process that none of the test items have a conditional probability greater than 2/3 at the standard. However, some of the panelists estimate the second item in the OIB as having a probability of greater than 2/3 (.67) for the person matching $\mathbf{minD_C}$ and others believe the item has a probability below that threshold value. Panelists 1 and 3 will place their bookmark on the first item in the OIB, while Panelists 2, 4, and 5 will go on to look at the third item in the OIB to determine if it has an estimated probability less than 2/3. Because all three of these panelists estimate the third item as having a probability less than 2/3, they will place their bookmarks on the second item.

TABLE 5.1

Estimated Probabilities for the Five-Item Example for Five Hypothetical Panelists without Guessing Effect

Item Number	Probability from IRT Model	Panelist Estimated Probabilities				
		1	2	3	4	5
3	.91	.71	.95	.93	.81	.89
1	.62	.65	.79	.58	.68	.74
4	.42	.40	.51	.33	.44	.43
5	.25	.23	.28	.30	.19	.38
2	.13	.14	.15	.15	.15	.11

An interesting observation from the information provided in Table 5.1 is that the Bookmark kernel is fairly robust to errors for items that have probabilities of correct response for persons at $\mathbf{minD_C}$ that are far from the mapping probability of 2/3. There can be substantial error in estimating the probabilities: for example, an estimate of .71 versus .91 for Panelist 1, and it has no consequence for the placement of the bookmark. However, errors in estimating the probabilities when the probability related to $\mathbf{minD_C}$ is close to 2/3 are critical. The differences in estimated probabilities for Item 1 result in substantial differences in the estimated standard because of the gap in the locations of items in the OIB.

Based on the mapped locations for the first two items in the OIB, two panelists have estimates of the standard of –1.11 and three have estimates of –0.31. The estimate of the standard obtained from the full group of five panelists depends on how the individual estimates are combined. If the mean value of the estimates is used as the standard, the result is –0.63. If the median value is used, the result is –0.31. For this simple example, if it is assumed that the distribution of examinee performance is normal with mean 0.0 and standard deviation 1.0, the standard based on the mean would have 74% of examinees above the estimated standard while the standard based on the median would have 62% above it. That is quite a large difference that results only from the selection of the statistic to use to estimate the standard. The intended standard of –0.5 has 69% of the examinee distribution above it.

The results shown by this example are an exaggeration of what is likely to be observed from an actual standard setting activity. If there were more items in the OIB and if they were equally spaced along the continuum, the change in estimated standard with bookmark placement would be smaller. The use of a larger number of panelists can also stabilize the results. With large numbers of panelists and the same amount of error estimating the probabilities, about 37% would place their bookmark on the second item in the OIB and 63% would place their bookmark on the first item in the OIB. That would result in a mean estimate of the standard of –0.82 and a median estimate of –1.11—quite different than the results in the small example. The difference is due to the instability of results from small samples. This result argues for having standard setting panels of sufficient size to stabilize the estimates of the standard.

The other source of error in the use of the Bookmark kernel is error in the estimation of the IRT parameters for the items on the test. If the items are calibrated from data from the administration of the test to different samples of examinees, the estimated parameters will be somewhat different from each sample. The amount of variation depends on the size of the sample and the estimation procedure used by the calibration software. Discussion of IRT estimation procedures is beyond the scope of this book. Interested readers should review the documentation of the software used for calibration and the extensive literature on the topic.

The purpose here is to show how the error in estimation of parameters can influence the estimate of the standard when the Bookmark kernel is used. For this purpose, the error distribution for the *a* parameter is assumed to have a standard deviation of 0.20, for the *b* parameter, a standard deviation of 0.15, and for the *c* parameter a standard deviation of .05. These are based on the average values of the standard errors of the parameters reported by Wainer and Mislevy (2000) based on 2,500 examinees. To generate item parameters with error, the original item parameters used to produce the plots in Figure 5.2 were adjusted by adding a random normal number sampled from a distribution with mean 0.0 and the appropriate standard error. The original item parameters and those with estimation error or shown in Table 5.2.

The item parameters with error were then used to generate the ICCs so they can be compared to the original results in Figures 5.2 and 5.3. Figure 5.4 shows the ICCs for the five items based on the item parameters with error. This plot shows the impact of error when compared to Figure 5.2. Note that the ordering of the items along the .67 line is slightly different from that for the item parameters without error. Items 5 and 2 have changed order. Item 5 is now estimated to be the most difficult item in the OIB where in actuality Item 2 is the most difficult item. This change has no consequence for the application of the Bookmark kernel in this example because those two items have locations that are not near the intended cut score.

Applying the Bookmark kernel to the OIB that has the item parameter estimation error included still has the bookmark placed on Item 1. However, because of the error in the item parameters for Item 1, the estimated standard is now –.73 instead of –.59 when no error was present in the parameters. Because of the error in parameters, the estimated standard is far from the intended standard. This will not always be the case because of the randomness of error. It could happen that the estimate with error would be closer to the intended standard, but that can never be known, because the intended standard is never known. All that can be concluded with certainty is that error in estimating item parameters will influence the estimate of the cut score.

TABLE 5.2

Original Item Parameters and Corresponding Parameters with Error

	a Parameter		*b* Parameter		*c* Parameter	
Item Number	Original	With Error	Original	With Error	Original	With Error
1	.69	.80	–.90	–1.09	.20	.13
2	.91	1.28	.73	.67	.09	.25
3	1.62	1.17	–1.36	–1.31	.20	.23
4	.92	1.09	–.30	.24	.12	.12
5	.79	.86	.32	.73	.06	.10

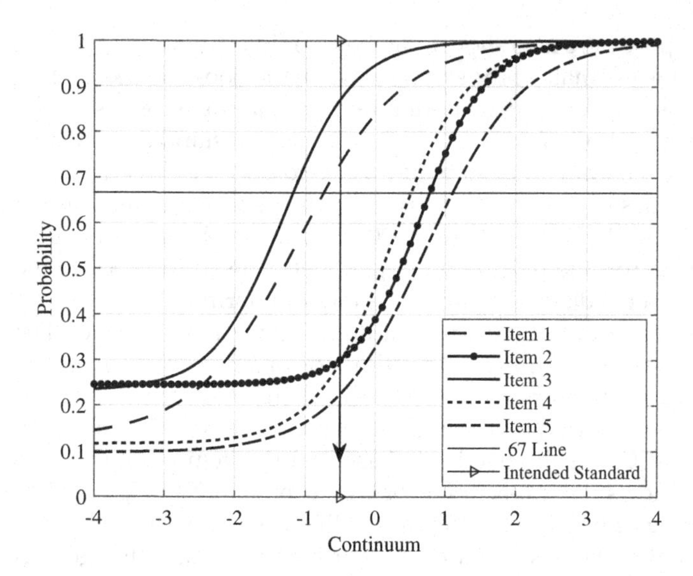

FIGURE 5.4
Item characteristics for the five-item test using item parameters with error.

Implementing the Bookmark kernel as originally designed includes the removal of the influence of the c parameters from the items when the OIB is created. Figure 5.5 shows the curves for the items with parameter estimation error when adjusted for the level of the c parameter. For this example, the adjusting for the influence of the c parameters did not change the ordering of the items in the OIB. However, the adjustment for c parameters did influence the estimate of the cut score. As with the unadjusted ICCs with error in the parameter estimates, those adjusted for c parameters have the bookmark placed on the second item in the OIB. That was Item 1 in the original test ordering. The location of that item using the 2/3 (.67) mapping criterion is –.53, a value close to the intended standard of –.50.

This example shows that although the Bookmark kernel seems to be an easily understood method that is simple to implement, there are many technical details that can have an impact on the estimated standard. If the persons who are designing the standard setting process are aware of the influence of the details, they can design a process that minimizes the effects. For example, the five-item test considered here has a large gap in location between the second and third items in the OIB when there was error in the parameter estimates. That means that moving the bookmark one item later would have a large impact on the estimated standard. This argues for making sure that there are enough test items in the OIB and that they are approximately equally spaced in location so that no large gaps occur.

The example given here attempted to implement the Bookmark kernel as Lewis et al. (1998) originally described it. However, the Bookmark kernel is often implemented in slightly different ways that have an impact

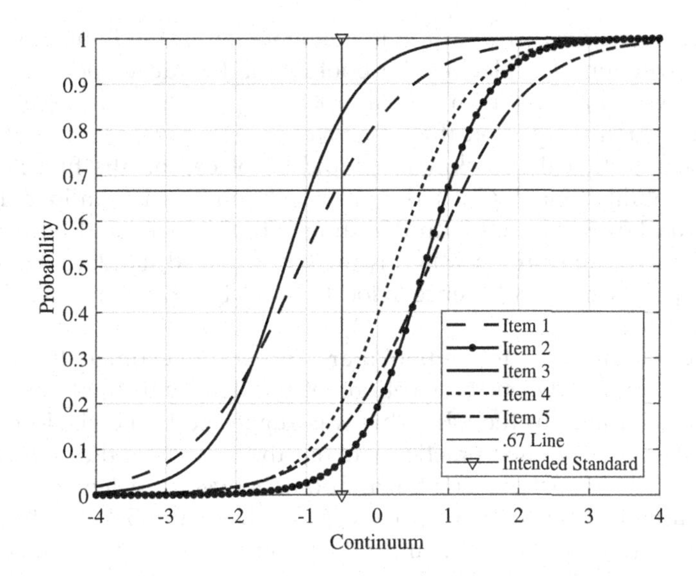

FIGURE 5.5
Item characteristics for the five-item test using item parameters with error and corrected for guessing.

on the estimated standard. The most critical is the mapping criterion for the items. Here 2/3 was used as the mapping criterion as suggested by the original authors of the method. But it has also been implemented using .5 (Wang, 2003) or .8 and there is some research that shows that panelists place their bookmark in the same place in studies when different mapping criteria were used. Hambleton and Pitoniak (2006) summarize the influence of the mapping probability on the resulting estimated standard. In general, a higher mapping probability results in a higher estimated standard. They also state that there is no theoretical justification for a particular mapping probability.

The grid lines in Figure 5.5 can be used to show the impact of keeping the bookmark on the same item when the mapping probability has different values. If .5 had been used as the mapping criterion, the estimated standard would be less than –1. If .8 were used for the mapping criterion, the estimated standard would be at approximately 0.0. In theory, the panelists should change their bookmark placements depending on the mapping criterion—they should place the bookmark on the first item in the booklet if the .8 mapping criterion was used. But panelists are reluctant to place a bookmark on the first item in the OIB, and they do not have a clear understanding of the influence of the mapping criterion. A good approach to this issue would be to research the probability that corresponds to the meaning of "can do" for the panelists and use that probability for the mapping criterion. That would make the understanding of the panelists and the implementation of the Bookmark kernel consistent.

A more subtle difference in the implementation of the Bookmark kernel is the instructions for placing the bookmark. The developers tell panelists to place it in the last item with a probability of correct response for the minimally qualified person that is above .67. However, Cizek and Bunch (2007) indicate that the bookmark should be placed on the first item that has a probability below .67 for the minimally qualified examinee. If these two approaches were used with the same OIB, the Cizek and Bunch (2007) instructions would result in a slightly higher estimated standard. How much higher depends on the difference in locations for the two items on either side of the standard.

Another modification to the Bookmark kernel is to estimate the standard as the midpoint between the locations of the item with the bookmark on it and the next item in the OIB. This was suggested by Lewis, Mitzel, and Green (1996), but common practice is to use the location of the item with the bookmark. The impact of using the midpoint between the item locations can be determined using the item curves shown in Figure 5.3. The bookmark should be placed on the first item in the OIB that yields an estimated standard of –1.1. But if the midpoint between the locations of the first and second items in the OIB is used to estimate the standard, the estimate is approximately –0.7, much closer to the intended standard of –0.5.

If the last item that is above the mapping criterion is used to estimate the standard, there tends to be a negative bias in the estimate of the standard because the location of the item must be below the standard unless it is exactly at the location of the standard. However, if the midpoint between the items located above and below the standard is used as the estimate, the estimates can be both above and below the intended standard removing the potential negative statistical bias.

There are many other variations in the details for the implementation of the Bookmark kernel, as there are for all the other kernels of standard setting methods. For example, panelists could be asked to place two bookmarks for a standard, one for the first item that is below the mapping criterion. Then they can be told to look further in the OIB to see if there are any other items that might be above the mapping criterion for the minimally qualified person. If so, they can put a second bookmark on the last item that is above the mapping criterion. The estimated standard is the midpoint between the two bookmarks.

Because there are so many variations for each standard setting kernel, it is important that the persons responsible for selecting the method do some analyses like those provided in the examples given here to show that the kernel can give estimates of the standard that are consistent with the intended standard. Of course, the intended standard is not known, so it is useful to try to recover several different intended standards in the range of the reporting score scale where the standard is likely to occur. Reckase (2006) shows this type of analysis for some implementations of the Bookmark and Angoff kernels.

5.2.2 The Angoff Kernel

The Angoff kernel (Angoff, 1971) is a method that asks panelists to review every test item on a test and estimate the probability that a person with characteristics described by $\mathbf{minD}_C$ would correctly answer the test item. That is, for the Angoff kernel, the **R** in Equation 5.1 can be expressed as $\hat{P}(u_i = 1|\mathbf{minD}_C)$. There are many different implementations of the Angoff kernel, two of which were described in Angoff (1971). One version is that the estimated probability is rounded to the nearest whole number, 0 or 1. This indicates whether the panelists estimate that the person with $\mathbf{minD}_C$ would answer the question correctly (1) or not (0). This has been labeled as the "yes/no method" by Plake and Cizek (2012). The other version is that the panelists report their probability estimates for each test item, often to two decimal places. Reporting the estimated probabilities is the version that is now most often labeled as the "Angoff method."

There is a subtle issue about how panelists are instructed to estimate the probabilities when using an Angoff kernel. Hambleton (1998) describes the instructions as: "your task is to specify the probability that this borderline student should answer each item in the assessment correctly," or "think of 100 borderline students, and then estimate the number of these borderline students who should answer an item correctly" (p. 95). Plake and Cizek (2012) argue that the borderline description ($\mathbf{minD}_C$) indicates what skills and knowledge that a person at the standard "should" have so the estimated probabilities are for how such a person "would" perform on the test items. The approach taken here is consistent with the position taken by Plake and Cizek (2012).

The estimate of the standard using the Angoff kernel is based on the statistical notion that the mean or expected value of a variable is equal to the sum of the values of the variable times the probability that they will occur. For example, if a test item is scored 1 for a correct response and 0 for an incorrect response, then the mean performance on the item for persons described by $\mathbf{minD}_C$ is given by

$$0 \times \hat{P}(u_i = 0|\mathbf{minD}_C) + 1 \times \hat{P}(u_i = 1|\mathbf{minD}_C) = \hat{P}(u_i = 1|\mathbf{minD}_C), \tag{5.4}$$

where i is the item number and $\hat{P}$ is the probability of the item score for an examinee with characteristics $\mathbf{minD}_C$ estimated by a panelist. Because the first term is multiplied by the 0 score, the expression simplifies to the estimated probability of a correct response. For the case of items scored 0 or 1, the estimated mean score on the full test from a panelist's estimates of probabilities on each item is the sum of the probabilities of correct responses,

$$\hat{\zeta} = \sum_i \hat{P}(u_i = 1|\mathbf{minD}_C). \tag{5.5}$$

This estimate of the standard is on the number-correct score scale. That is, it gives the expected number of correct responses for persons with characteristics $\mathbf{minD_C}$.

The five-item test used as an example for the Bookmark kernel can also be used as an example for the Angoff kernel. However, when the Angoff kernel is used, panelists usually discuss the influence of guessing and include that influence in their estimates. If a panelist perfectly understands $\mathbf{minD_C}$ and the functioning of the test items, and makes probability estimates without error, and if the IRT model accurately represents how the examinees interact with test items, then the panelist would give probability estimates for the items that are in column 2 of Table 5.3. This is the probability computed from the IRT model for examinees at the intended cut score. In this case, the IRT model is assumed to accurately represent the probability of correct response and examinees' locations on the hypothetical continuum. The estimate of the standard on the number-correct score scale would then be the sum of the values in that column, 2.62 out of a maximum of 5 for the example. Comparing this value to the intended standard of –0.5 requires transforming it to the scale defined by the IRT model. This is done using the test characteristic curve (TCC) that shows the relationship between the IRT θ-scale and the number-correct score scale. The TCC is obtained by summing the ICCs for the items at each θ-value. Figure 5.6 shows the TCC for the five-item test example.

For the case when a panelist accurately estimates the probabilities of correct response for the examinees characterized by $\mathbf{minD_C}$, the recovery of the intended standard is very near the value of –0.5. The arrows in Figure 5.6 show the mapping of the probability to the estimated standard. The estimated number-correct score is mapped through the TCC to the θ-scale. The only difference between the estimated standard and the intended one is due to rounding error resulting from giving the ratings to two decimal places. However, panelists are not able to make such precise estimates of these probabilities (Plake & Cizek, 2012). To show the influence of errors in estimation of the probabilities on the estimation of the standard, the same error

TABLE 5.3

Estimated Probabilities for the Five-Item Example for Five Hypothetical Panelists

Item Number	Probability from IRT Model	Panelist Estimated Probabilities				
		1	2	3	4	5
3	.93	.88	.94	.68	.98	.92
1	.69	.71	.59	.76	.59	.56
4	.49	.38	.54	.35	.53	.34
5	.29	.36	.23	.44	.32	.28
2	.21	.24	.25	.19	.24	.36
Sum	2.62	2.57	2.55	2.42	2.66	2.46

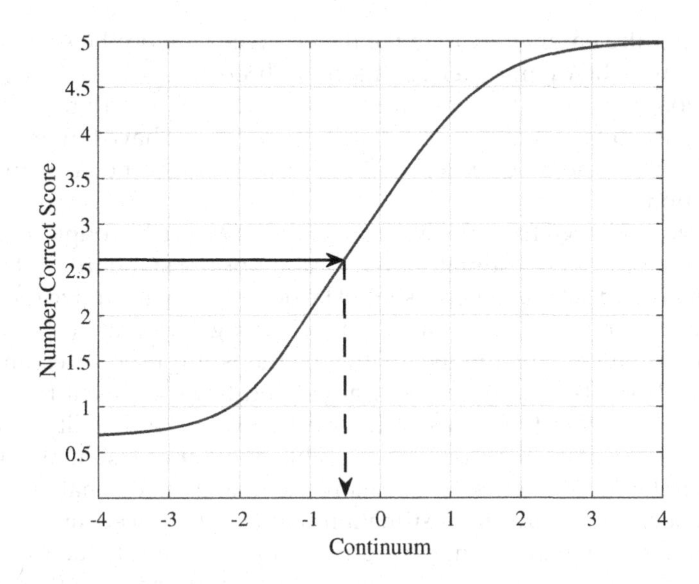

FIGURE 5.6
Test characteristic curve (TCC) for the five-item test with mapping of estimated number-correct score to the θ-scale.

model as was used for the Bookmark kernel was applied to the probabilities in column 2 of Table 5.3. The estimated probabilities with error are shown in columns 3–7 for five hypothetical panelists. The sums of the probabilities in each column are given in the last row of the table. The values for each of the simulated panelists shows the influence of errors in estimating the probabilities on the standards estimated on the number correct score scale. While there are slight differences in the estimated standards, all indicate that a score of 2 on the five-item test is below the standard, and a score of 3 is above the standard.

When the estimates from the five panelists are converted to the θ-scale by mapping them using the TCC, the results are –0.55, –0.56, –0.67, –0.46, and –0.64, respectively. The average of the five estimates is –.58. This is a slight underestimate of the intended standard, –.5. However, this result assumes that the parameters for the items have been accurately estimated. If the parameter estimates with error listed in Table 5.2 are used to get the estimates, the results are, respectively, –0.36, –0.38, –0.50, –0.27, and –0.47 for the five panelists with a mean value of –.40. The estimated standard using the probabilities estimated without the error in individual judgments but using the item parameters with error is –.31.

Comparing the estimates of the intended standard obtained without error, with judgment error only, with parameter estimation error only, and with both types of error is difficult because the IRT scales used for the comparisons are not the same. Certainly, it is reasonable to compare the estimate from judgment error, –0.58, to the intended standard, –0.5, because both assume that the parameters to make the estimates are on the same θ-scale. But the scale when parameter estimates with error are not the same. The distribution of θ-estimates for examinees would change with the change in the item parameter estimates. The distributions would have larger standard deviations than the distribution without error because the error increases the variation.

The major message from the example is that there are multiple sources of error in the process of estimating the location of the standard on the scale and the standard setting process should be designed to minimize the impact of the sources of error. This can be done through accurate estimation of the item parameters for the test, if IRT methods are used to estimate the standard. Panelists should be well trained and have sufficient feedback to help them minimize the errors in their estimates of probability of correct response. Also, if panelists rate a reasonable number of test items, the estimate from the Angoff kernel will be more accurate than for smaller numbers of items, because the errors of estimation will tend to cancel out.

As with the Bookmark kernel, there are many variations to the Angoff kernel. It has become common to label them all as "modified Angoff" methods. This label does not have much meaning because there are so many modifications that a detailed description of the standard setting process is needed to distinguish among them. Most of the modifications are designed to reduce the judgment load for panelists. For example, panelists may be asked to make probability estimates to only one decimal place, or into ranges such as 0–.19, .20–.39, .40–.59, .60–.79, and .80–1.00. As indicated in the introduction to this section, Angoff (1971) suggested that panelists estimate whether the examinee described by $\mathbf{minD}_C$ will answer the test items correctly or incorrectly. This is essentially the same as having them round their probability estimates to 0 or 1.

The information in Table 5.3 can be used to show the impact of rounding the probability of correct response values to 0 or 1. If all the values above .5 are rounded to 1 and those below .5 are rounded to 0, then the ratings from the IRT model and the panelists with error in their ratings are shown in Table 5.4.

TABLE 5.4

Estimated 0, 1 Ratings for the Five-Item Example for Five Hypothetical Panelists

Item Number	Based on Probability from IRT Model	Panelist Estimated Item Scores				
		1	2	3	4	5
3	1	1	1	1	1	1
1	1	1	1	1	1	1
4	0	0	1	0	1	0
5	0	0	0	0	0	0
2	0	0	0	0	0	0
Sum	2	2	3	2	3	2

First, it should be noted that the ratings and estimated cut score based on the probabilities from the IRT model yield two correct responses and three incorrect responses, yielding an estimated standard of 2, a value that is lower than the 2.62 that would be obtained if the actual probabilities had been used to compute the standard. Translating the value of 2 to the IRT scale yields a value of approximately –1 (see Figure 5.6). This is substantially lower than the intended standard of –0.5.

When panelist error is added to the probabilities before rounding to 0 or 1, the estimated standards are either 2 or 3, with an average of 2.4. The transformation of this value to the IRT scale through the test characteristic curve is approximately –0.67. Using multiple panelists, even with errors in the ratings, yields a closer value to the intended standard, but it is still an underestimate of the intended standard of –0.5.

The reason for the underestimation of the standard is a result of the difficulty level of the test. Suppose that a test is very difficult, and most items have a predicted probability of a correct response at the intended standard of less than .5. All those items will be rounded to 0 when the Yes/No kernel is used, and they will not contribute to the sum that is used to estimate the standard. The few items that have probabilities above .5 will be rounded up to 1, but the gains made from rounding up do not balance those from rounding down. An opposite effect occurs if the test is very easy. Most items will be rounded up to 1 with few rounded down to 0, resulting in an estimated standard that is higher than the intended standard.

A 50-item simulated test was used to show the cumulative effect of the rounding to 0 or 1 and how the rounding effect interacts with test difficulty. Table 5.5 provides the means and standard deviations of the item parameters for the 3PL IRT model used to generate the data for the test. The summary of the item parameters shows that the test is well targeted to an intended standard at –0.5 on the IRT scale and that there is a reasonable spread of difficulty of the items.

The IRT model and the item parameters were used to compute the probability of a correct response for each item at the intended standard. These probabilities would be the equivalent of values estimated by panelists if they performed the task required by the Angoff kernel without error, assuming the IRT model was an accurate representation of the interaction between

TABLE 5.5

Means and Standard Deviations of the 3PL Model Parameters Used to Generate the Item Response Data

	Item Parameter		
Statistic	*A*	*b*	*c*
Mean	.96	–.44	.16
Standard Deviation	.34	.91	.07

examinees and the test items. The number correct score that corresponds to the intended standard of –.5 under these ideal conditions is 27.99 out of a maximum score of 50. If that value is mapped through the TCC for the test, it recovers the value of –.5 on the IRT scale. If each of the probabilities is rounded to 0 or 1, that is, the Yes/No kernel is used, the estimated standard is 29, slightly higher than the value that is consistent with the intended standard. There is not much difference between the two values in this case because the test is of medium difficulty and the intended standard is near the middle of the IRT score range.

To show the difference in results from the use of the two different kernels as the difficulty of the test changes, the difficult of the test was shifted by adding a constant to all the *b* parameters. If the constant has a negative value, say –3, the shift in difficulty makes the test much easier. If the constant is a positive value, say 3, the test is much more difficult. To show the influence of test difficulty on the standards estimated using the two kernels, shifts were made from –3 to 3 in increments of 0.1. For each of the resulting simulated tests, the standard corresponding to the intended standard of –0.5 on the IRT scale was determined from either the sum of the probability of correct responses for the test items assuming the intended standard or the sum of the probabilities rounded to 0 or 1. Figure 5.7 provides a summary of the results.

The results of this example show that there are large differences in the estimated standard from the two kernels as the test becomes easier or more difficult. The value of 0 on the Adjustment-to-Difficulty scale was the original

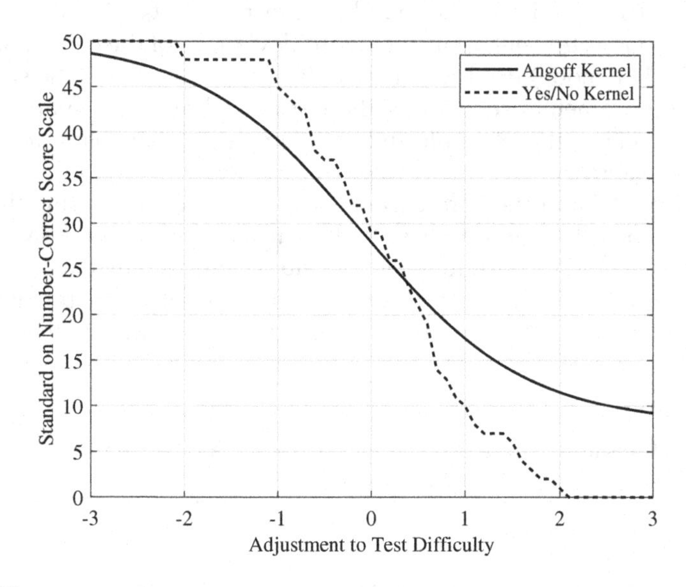

FIGURE 5.7
Estimated standards on the number-correct score scale for the Angoff and Yes/No kernels for different amounts of shift of test difficulty.

result without any adjustment. For that case, there was little difference between the two kernels. However, as the difficulty of the test became more extreme, the differences between the two kernels were greater. Making the test easier by one unit (adding –1 to the *b* parameters) results in a standard of 38.17 from the Angoff kernel and a standard of 44 from the Yes/No kernel if panelists make judgments without error. The differences are more extreme as the test becomes more difficult. If the *b* parameters are increased by one unit on the IRT scale, the resulting standards are 17.41 for the Angoff kernel and 10 for the Yes/No kernel. These differences would result in major differences in the proportions above the standard for these tests.

This type of simulation is useful for determining if there are technical characteristics of a kernel that tend to produce statistical bias in the estimated standard assuming that panelists make judgments that are logically consistent with the method. These characteristics of the kernel can be considered when selecting a method and designing the standard setting process. However, it is unlikely that panelists functioning exactly as the simulation suggests they should. Work by Wyse (2018) indicates that panelists tend to regress their probability estimates toward the center of the probability scale. This type of estimation error could be modeled by simulation as well. The goal should be to evaluate the functioning of a kernel in as realistic way as possible. The design of such studies is described in more detail at the end of this chapter.

5.2.3 The Body of Work Kernel

The Body of Work kernel was developed in the 1990s for use with the surge in assessments that used constructed-response test items such as writing samples, science projects, etc. It was originally described by Kahl, Crockett, DePascale, and Rindfleisch (1994), but there are now numerous descriptions of the method (Cizek & Bunch, 2007; Hambleton & Pitoniak, 2006; Kingston & Tiemann, 2012; Pitoniak & Cizek, 2016). The basic task for panelists when using this kernel is to classify actual examples of student products into categories above or below the intended standard. If there are multiple cut scores, then the student products are sorted into the number of categories defined by the cut scores. The number of categories is the number of cut scores plus one. In the framework used here to describe the translation process, panelists are asked to determine whether a sample of work gives evidence of being above the requirements given by $\mathbf{minD}_C$. If so, they classify the work as being above the standard. Otherwise, it is considered below the standard. The information, **R**, collected from the panelists are the classifications of the work samples and the scores given to those work samples obtained through an independent scoring process. The scores for the work sample are not provided to panelists when they are doing the classification. The name for the kernel, Body of Work, comes from the material that panelists review for each examinee when making the classification.

As with the other kernels in this chapter, the Body of Work kernel assumes that there is a policy definition, $\mathbf{D_P}$, and that it has been translated into content-specific definitions, $\mathbf{D_C}$, for the regions on the continuum defined by the hypothetical construct, ζ, that are defined by the intended standard or standards on that continuum. The original development of this method used scores based on the sum of scores on a series of performance tasks that made up the test as the reporting score scale for the standard setting. For that reason, the intended standard implied by $\mathbf{D_P}$ and $\mathbf{D_C}$ is indicated on the true score scale, τ_{Ag}. It is also possible to use IRT models to define the reporting continuum for tests of this type.

As with other kernels, the Body of Work kernel requires that materials be organized in a certain way before the standard setting process begins. For this kernel, the work samples produced by examinees are scored and a total score is produced for each "body of work." Kingston and Tiemann (2012) indicate that they had all work samples independently scored by two individuals to increase the accuracy of the scores used in the process. Then, the work samples are sorted into sets based on the total summed score for the work samples. The sets based on single scores were then grouped into folders for score ranges. Kingston and Tiemann (2012) use four-point score ranges. For a test with a maximum summed score of 50, they used nine score ranges starting with a score of 12. From these, six work samples were selected that covered the observed score range to use for training; two from the lowest score in the range and one from the highest score in the range were selected from each folder for a "range finding" set. After an initial estimate of the location of the standard was found, one or two folders of work samples near the estimated summed score were selected for "pinpointing." Each of these folders contained 20 work samples. The panelists classified the pinpointing work samples into categories as consistent or higher than the $\mathbf{D_C}$ or less than the $\mathbf{D_C}$. Finally, logistic regression was used with the classifications and the observed scores for the work samples to identify the score level that had a .5 probability of correctly classifying examinees as matching the requirements of $\mathbf{D_C}$.

Before considering the details of the Body of Work kernel, it is useful to evaluate how well it would work if the scoring of the work samples were perfectly reliable and panelists made the classifications with perfect accuracy. A basic requirement for a good standard setting kernel is that it can recover the intended standard when the tasks defined by the kernel are performed exactly as specified. As with other kernels, the Body of Work kernel assumes that there is a $\mathbf{D_P}$ that implies a continuum of performance. In this case, the continuum of performance is operationalized by the scores obtained for the various entries in the work samples for an individual. If these work samples are perfectly scored on the construct of interest, the scores assigned to each are the true scores, τ_i, for the persons who produced the work samples. That means, when the work samples are ordered to create the range-finding folders, they are ordered perfectly according to the construct of interest, and the assigned scores perfectly match the scores on the hypothetical construct that defines the continuum. Furthermore, if there is a well-written $\mathbf{D_C}$, then the panelists can make a clear connection between the work in the range-finding folders and $\mathbf{D_C}$. This will allow them to make a perfect classification of work samples as meeting or exceeding the requirements of $\mathbf{D_C}$, or not meeting those requirements. That means that for the observed score scale, say from 0 to 50, they can identify a score, say 28, that has work samples that meet the requirements of $\mathbf{D_C}$ and all those above 28 exceed those requirements. The work samples corresponding to 27 and below will be deficient in some way when compared to the requirements of $\mathbf{D_C}$. In this case, the estimated standard is 28 and it exactly matches the intended standard implied by $\mathbf{D_P}$ and $\mathbf{D_C}$.

The above example is an interesting thought experiment, but, of course, the Body of Work kernel will never work that way in practice. $\mathbf{D_C}$ will not have the detail and clarity that is needed, the scoring of the work samples will not be perfectly reliable, and there will be errors in classification when the panelists compare the work samples to the requirements of $\mathbf{D_C}$. The important question to be considered is how well the Body of Work kernel recovers the intended standard when it is applied in a realistic situation?

When checking the functioning of the Body of Work kernel, it will be assumed that the selection of entries in the work samples is consistent with the hypothetical construct implied by the $\mathbf{D_P}$. If that is not true, there are serious issues with the validity argument for the assessment that are beyond the scope of a discussion of standard setting. Therefore, it is assumed that the scores from the scoring of the work samples are a reasonable operationalization of the hypothetical construct. However, the accurate and consistent scoring of open-ended assessment tasks is challenging. In this case, following Kingston and Tiemann (2012), it is assumed that there are several entries in the work samples and each of those entries are scored by two independent well-trained individuals. The overall score for the work sample for an individual is the sum of the scores on each of the entries. For the example used here, the reliability of the reported score on the work sample is assumed to be .80, a value that is consistent with the state-of-the-art for scoring open-ended tasks. Furthermore, the score scale ranges from 0 to 50 with the mean for a sample of 1,000 examinees of 32.63 and a standard deviation of 9.72. The standard error of measurement for the reported scores is 4.35. When modeling the data, the observed scores are assumed to be randomly sampled from a normal distribution with the mean equal to the true score τ_i and a standard deviation equal to the standard error of measurement. For each true score, the value sampled from the error distribution is added to the true score and then rounded to a whole number.

When Kingston and Tiemann (2012) implemented the Body of Work kernel, they divided the score range from 0 to 50 into intervals of four score points, starting with the score of 12. That is, the first interval included 12, 13, 14, 15. They skipped the lower scores because they had low frequency of

TABLE 5.6

Observed Scores for Work Samples Selected for Range Finding and the True Scores Used to Generate Them

Folder	Observed Score	True Score	Folder	Observed Score	True Score	Folder	Observed Score	True Score
1	12	12	4	24	21	7	36	27
	12	14		24	22		36	34
	15	20		27	34		39	44
2	16	13	5	28	24	8	40	39
	16	18		28	25		40	43
	19	23		31	31		43	49
3	20	22	6	32	28	9	44	41
	20	23		32	34		44	41
	23	29		35	39		50	50

occurrence, and the expectation was that those scores would have a lot of error. From each four-score range, they selected two low scores and one high score to use for the range-finding set of 27 work samples. The work samples for each range were included in a folder resulting in nine folders. Of course, these selections were made using observed scores. The generation of data for the other folders was done in the same way. Table 5.6 presents the observed scores that were selected for each folder and the true scores that were used to generate the observed scores. Even though there are notable differences between the true scores and the observed scores, the pairs of scores are highly correlated at a value of .94. The magnitude of this correlation is partially due to the essentially uniform distribution of observed scores. Based on the magnitude of the reliability alone, the correlation would be expected to be .89, assuming a normal distribution of true and observed scores.

The panelists involved in the Body of Work standard setting process are not given the observed scores for the work samples. Instead, they are given some minimal training on the scoring rubrics for the entries in the work sample and they take the test themselves. Then they score their own work to get a sense of the scoring process. However, they are not trained in scoring in the same way as those who do operational scoring. Furthermore, they are told to consider "overall quality" of the work sample when ranking them and classifying them as matching or not matching the requirements of the $\mathbf{D}_C$. As a result, reliability of judgments by the panelists about the performance represented by the work samples is likely less than the reliability obtained for the operational scoring. For this simulation, it is assumed that the reliability of the evaluation by panelists is .6.

The Body of Work process described by Kingston and Tiemann (2012) used 16 panelists. To model this, 16 values were sampled from the distribution specified by the standard error of measurement (6.15) from a reliability of .6. Separate values were sampled for each panelist and the values were added to the true scores for the work samples used in the range-finding activity. Then the values were compared to an intended standard of 28 to determine if the panelists' judgments were that the work sample matched $\mathbf{D}_C$ or not. The proportion defined as matching $\mathbf{D}_C$ or higher was tallied across simulated panelists. Table 5.7 presents the proportions along with the folder number and observed scores.

Some of the values in the table seem anomalous when compared to adjacent values. For example, the last entry in Folder 4 (.8125) seems very high compared to the other values in Folders 4 and 5. The reason for this is that the observed score is an underestimate of the true score that was 34. When classifying work samples, the panelists do not see the observed score, so they are making classifications on wholistic judgments of the entries in the folder. These anomalies result from a combination of the error in the formal scoring process and in the judgments of the panelists.

After collecting range finding classifications, Kingston and Tiemann (2012) recommend finding folders that meet the following two criteria:

> (a) one-third or more of the panelists thought at least one paper [work sample] in the folder belonged to the performance level; and (b) two-thirds or more of the panelists did not think any of the papers in the folder should be in the performance level.
>
> **(p. 208).**

In this case, Folders 2 through 9 all meet the first condition. Each of those folders have at least one work sample that simulated panelists selected over one-third of the time to match $\mathbf{D}_C$. Folders 1, 2, and 3 have .98, .79, and .77 proportion of the panelists not thinking any of the work samples matched the requirements of $\mathbf{D}_C$. Also, Folders 4 and 5 were close to the second criterion, with .62 and .65 simulated panelists not classifying work samples as matching the requirements of $\mathbf{D}_C$. It is not clear how rigidly the rules should be followed, but for this example, it is assumed that Folders 2 and 3, that match both criteria, are used for the next part of the implementation of the kernel, the pinpointing round.

For the pinpointing round, 20 additional work samples were selected from the folders meeting the criteria and panelists are asked to classify those on their match to $\mathbf{D}_C$. In this case, there would be 20 work samples from each of Folders 2 and 3, yielding 40 additional work samples for the classification task. In this case, only 37 simulated work sample were present in the data generated for the example, so those were used for the pinpointing round. For each of the 37 work samples, the classifications by the 16 simulated panelists were determined in the same way as for range finding. Then, for each score point in the pinpointing folders, the proportion of work samples matching the $\mathbf{D}_C$ was computed. Table 5.8 shows the results. The table contains only a limited range of observed scores because those were the only ones included in the selected pinpointing folders.

TABLE 5.7

Proportion Classified as Matching D_C for Each Work Sample and the Observed Scores Used to Generate Them

Folder	Observed Score	Proportion Matching D_C	Folder	Observed Score	Proportion Matching D_C	Folder	Observed Score	Proportion Matching D_C
1	12	0	4	24	.1250	7	36	.5625
	12	0		24	.1875		36	1.0
	15	.0625		27	.8125		39	1.0
2	16	.0625	5	28	.2500	8	40	1.0
	16	.1250		28	.2500		40	1.0
	19	.4375		31	.5625		43	1.0
3	20	.1250	6	32	.4375	9	44	1.0
	20	.1875		32	.8125		44	1.0
	23	.3750		35	1.0		50	1.0

TABLE 5.8

Proportion Classified as Matching $\mathbf{D_C}$ for Each Work Sample in the Pinpointing Folders

	Folder							
	2				3			
Observed Score	16	17	18	19	20	21	22	23
Number of Work Samples	4	7	4	4	5	7	3	3
Proportion Matching $\mathbf{D_C}$	.09	.10	.05	.12	.34	.19	.15	.40

The observed proportions matching $\mathbf{D_C}$ do not increase with much regularity as the observed scores increase. The correlation between the observed scores and proportions is only .36, although it is statistically significant. Part of the variation is due to the small number of work samples for each observed score point. For an observed score of 23, there were only 3 work samples yielding 3 × 16 = 48 observations going into the computation of the proportion. Part of the variation is due to the presence of error in both the observed scores and the classifications by the simulated panelists. Because this is simulated data, it is possible to compute the correlation between the proportions and true scores for the work samples. That correlation is .83. It is much higher because it is based on 16 simulated panelists and the error in the observed scores does not influence the results. This suggests that when using the Body of Work kernel, it is important to get reliable scores for the work samples because error in those scores has a strong impact on the results.

After the pinpointing data are collected, Kingston and Tiemann (2012) recommend analyzing the relationship of the proportion matching $\mathbf{D_C}$ and the observed scores using logistic regression. In this case, the MATLAB routine "fitglm" was used to estimate the logistic regression function used to predict the logit of the proportions from the observed scores. The result of the analysis was the following equation:

$$\ln\left(\frac{p}{1-p}\right) = 0.25089\,\text{score} - 6.6143. \tag{5.6}$$

Figure 5.8 shows the relationship between the proportion of the work samples judged to be above the standard and the observed scores for the work samples. The figure also shows the curve for the logistic function given in Equation 5.6. Note that the curves do not reach the proportion of .5 for any observed scores. This suggests that the rules for selecting the pinpointing work samples did not result in a set that was well suited to estimate the point where the probability of classification of the work samples above the standard is .5. However, Equation 5.6 can be solved for the score that would predict the value of .5 by setting the left side to 0, $\ln\left(0.5/(1-0.5)\right) = 0$, and

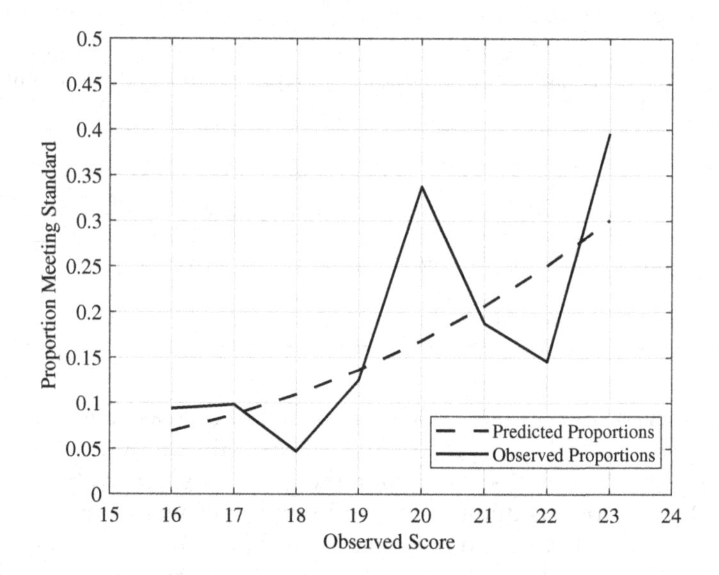

FIGURE 5.8
The observed relationship between the proportion classified as being above the standard and the observed score for the work samples along with the predicted curve from the logistic regression analysis.

solving for the score. The result is a score of 26.36, slightly lower than the intended value of the standard used in this simulation.

The results of this small simulation should not be over generalized. There is little research on the functioning of the Body of Work kernel under different conditions. It would be useful to know how well the kernel functions when intended standards are at different points on the reporting score scale. This example had the intended standard near the middle of the scale. It would be important to know if there are estimation problems when the intended standard is very high or very low. It would also be useful to have some comparisons of the functioning of the kernel with different numbers and distributions of work samples. However, the small simulation study reported here gives some support to an expectation that the use of the Body of Work kernel will yield an estimated standard that is in the vicinity of the intended standard.

5.2.4 The Contrasting Groups Kernel

The Contrasting Groups kernel (Livingston & Zieky, 1982) is based on the classification of individuals into groups that either meet the intended standard or do not. This task is usually done by persons who are well acquainted with the individuals such as classroom teachers classifying their students, or supervisors in a job setting. Independent of this classification, the target individuals are given the test that is the operationalization of the construct implied by $\mathbf{D}_P$. Then the point on the operational reporting score scale is estimated according to some statistical criteria. Livingston and Zieky (1982) suggest the score level where there are equal probabilities of passing and failing as the estimated standard, but other estimation techniques have been suggested. Livingston and Zieky (1982) also suggest smoothing the observed score distributions before estimating the standard.

Under the ideal conditions of knowing the true score for each examinee and panelists making judgments without error, there is no doubt that the Contrasting Groups kernel can accurately translate the intended standard described in $\mathbf{D}_P$ and $\mathbf{D}_C$ to a point on the reporting score scale. Under those conditions the Contrasting Groups kernel will precisely divide the true-score distribution into two parts above and below the cut score. For this reason, scholars such as Shepard, Glaser, Linn, and Bohrnstedt (1993) seem to consider it as a gold standard for standard setting (p. xxvi, 23). However, operationally the Contrasting Groups kernel uses observed test scores for the examinees, and there is error in the classification of examinees into groups. There are no rigorous studies of the accuracy of classifications when using the Contrasting Groups kernel, but it is not likely that the classifications are more accurate than the typical ratings of student performance by teachers.

To demonstrate the functioning of the Contrasting Groups kernel, the same level of test and panelist error as was used for the Body of Work kernel is used here. It is assumed that there is an intended standard as described in $\mathbf{D}_P$ and $\mathbf{D}_C$ and that panelists each classify 30 individuals as being above or below their understanding of the intended standard with a reliability of .60. It is also assumed that the test administered to the individuals has a reliability of .90. Furthermore, there are 20 panelists so that there are $20 \times 30 = 600$ observed scores that are the data that are available for estimation of the standard. The information **R** that is collected using the Contrasting Groups kernel is a set of 600 pairs composed of a classification above or below the standard and an observed test score for each individual. The goal is to compute an estimate of the point on the reporting score scale that corresponds to the intended standard.

Figure 5.9 shows the distributions of observed scores for the examinees classified as above the standard and those classified as below the standard when the intended standard was a true score of 28. That value was used for the example to allow comparison with the results from the Body of Work kernel. One finding from this example is the difference in frequencies for the two distributions. The distribution of examinees classified by the simulated panelists as below the cut score implied by $\mathbf{D}_P$ and $\mathbf{D}_C$ had 141 observations. The simulated results had most examinees, 459, above the standard.

Figure 5.10 shows the conditional proportions above the standard at each score point and the smoothed curve. Livingston and Zieky (1982) recommended estimating the standard at the point on the score scale where the proportion classified above the cut score is .5 for the smoothed curve. This is the point where the curves for the two distributions intersect. In this case,

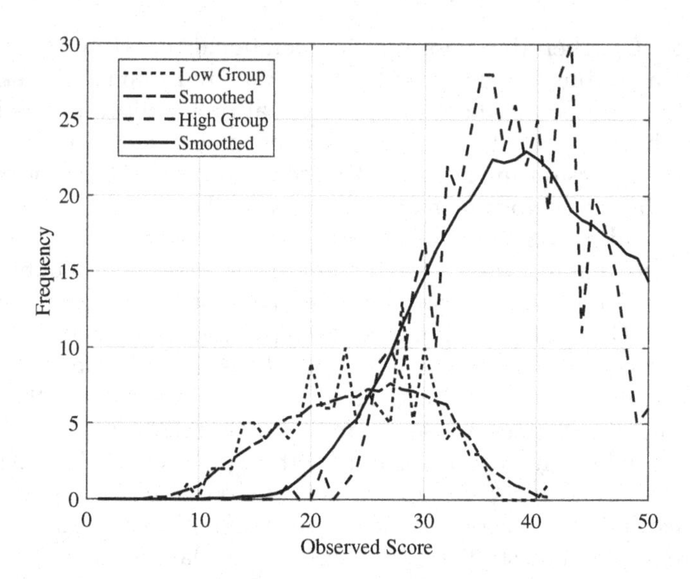

FIGURE 5.9
Distributions of observed scores for examinees classified as above and below the standard.

the estimate is 27.3, just below the intended standard of 28. Of course, this estimate depends on how the smoothing was done and the way that the standard was estimated. In this case, the MATLAB (The Mathworks, 2022) program "smoothdata" was used to smooth the curve using a moving average algorithm, and linear interpolation was used to estimate the cut score from the smoothed curve.

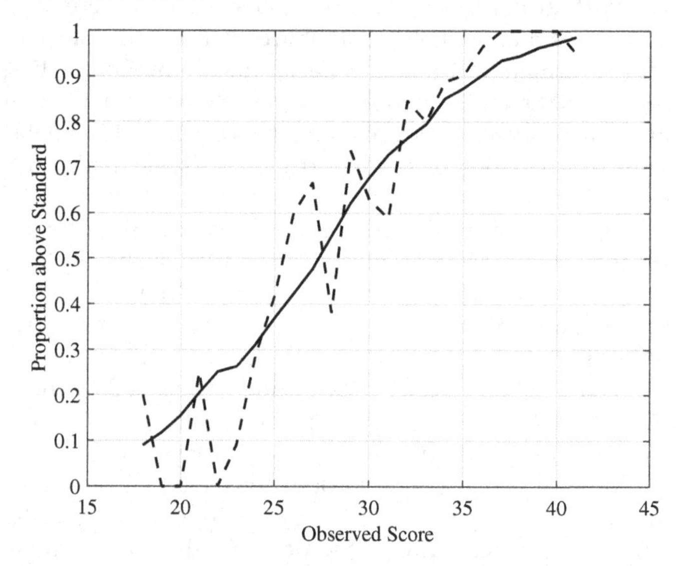

FIGURE 5.10
Proportion of simulated examinees classified as above standard by observed score.

5.3 Criteria for Selecting a Standard Setting Kernel

This chapter provided examples of only a small number of the possible kernels for standard setting procedures. Each of those described has many variations, and new kernels are being developed as assessment procedures change and as the understanding of the function of a standard setting process gets more nuanced. Given the large number of options available to those implementing a standard setting process, how can those in charge of developing the standard setting process choose among them? As much as it would be desirable to state here that a specific kernel is the best of those developed so far, that is not possible because there are too many variables in the way a kernel is implemented. One kernel might be the best option when there are highly knowledgeable and well-trained panelists with several days assigned to the process, while that same kernel might give poor results for different test content when the time allocated is one day. This means that it is the responsibility of the persons designing the process to provide information to support the particular way that their proposed kernel is implemented to show that it will yield defensible results. Each kernel has potential weaknesses. It is important that procedures for addressing those weaknesses be considered when that kernel is selected.

Two general criteria for selecting the kernel for the standard setting process are suggested here. Many more could be listed, but these two are important as support to the theoretical position taken in this book.

5.3.1 Criterion Number 1

The first criterion is a technical/statistical one. A minimal criterion for a standard setting kernel is that it be *able to provide data that can be used to accurately estimate the point on the reporting score scale that is the translation of* $\mathbf{D}_P$ *and* $\mathbf{D}_C$. A practical way to think about this criterion is whether a panelist who is knowledgeable and well trained and who follows the procedure specified by the kernel exactly as it is designed can recover the standard that they intend. Suppose a panelist knows what the point should be (this is like knowing the true score for a student on a test), can they perform the tasks required by the kernel to accurately recover that known point? Of course, that point is never known with certainty, but through simulation studies, the kernel can be checked to see if the data it produces, **R**, can be used to estimate hypothesized known points. The examples provided in this chapter used this methodology to check whether the different kernels could be used to estimate known standards. Reckase (2006) used this criterion for checking the capability of specific Angoff and Bookmark kernels for recovering intended standards spaced along the reporting score scale.

The analysis of the Bookmark kernel given in this chapter provides insights into this criterion. If the OIB has test items evenly spaced according to the mapping criterion and a panelist makes the bookmark placement exactly as specified by the Bookmark kernel, the intended standard can be recovered with a level of accuracy that is specified by the difference in difficulty of adjacent items in the OIB. However, if the items in the OIB have large distances between the mapped values of some items and clustering of difficulties for other items, it may be impossible to recover intended standards located in the parts of the reporting score scale where large distances between items occur. This implies that the design of the OIB is critical to the proper functioning of the Bookmark kernel. Of course, there are other important technical issues such as the accuracy of the item calibrations and the amount of error in the judgments made by panelists. Using large calibration samples and enough well-trained panelists to yield small standard errors of the estimated standard should address these issues.

The Angoff kernel is relatively unaffected by the characteristics of the items in the test if panelists can make accurate judgments. However, Wyse and Reckase (2012) and Wyse (2018) have shown that the recovery of intended standards can be influenced by rounding the probability estimates or by regression effects for those estimates. It was shown here that the most extreme rounding, the Yes/No kernel, can result in statistical bias that can have notable effects on the estimated standard. For the Yes/No kernel to accurately recover the intended standard, the intended standard must be near the middle of the distribution of difficulty of test items.

5.3.2 Criterion Number 2

The second criterion for selecting a standard setting kernel is the likelihood that well-trained panelists can implement the kernel as specified. In this regard, the Angoff kernel has received the most attention. Because of its use for NAEP standard setting, it has received very careful scrutiny and researchers such as Shepard, Glaser, Linn, and Bohrnstedt (1993) have indicated that the probability estimates required of the Angoff kernel are too cognitively demanding to be provided by panelists at a level of accuracy that is useful. The counterargument is that, like scores from dichotomous test items, individual probability estimates for items are not vary accurate, but if enough of them are aggregated, the estimates of standards can be sufficient for the proposed uses. That is one of the messages of the simulations given earlier in this chapter when substantial error was modeled for panelists' estimates of conditional probabilities for items. However, the generalizability of such results depends on how well the modeling of error matches the actual error distributions for an implementation of a kernel.

Some kernels, such as the Bookmark and Yes/No kernels, appear to be easier for panelists to implement because those kernels only require estimating whether a conditional probability is above or below a specific value rather then estimating the numerical value of the probability. On the other hand, the Contrasting Groups kernel appears to be simple, but it requires panelists to evaluate individuals on the set of skills and knowledge included in $\mathbf{D_P}$ and $\mathbf{D_C}$, while ignoring all other things they might know about the individuals. That would seem to be a very difficult task.

In all these cases, the ability of panelists to successfully provide the data specified by a kernel is dependent on the training and practice that they have had for making the judgments that are involved. The data obtained from the use of a kernel may have a lot of error when panelists receive 30 minutes of training but can have acceptably small amounts of error after two hours of training. We accept most of the judgments of sports referees and Olympic judges as they evaluate very complex situations because we believe they have been thoroughly trained to make those judgments. Even then, however, methods have been developed to check those judgments (i.e., instant replay) and retrain as necessary to get acceptable levels of accuracy. Unfortunately, such quality checks have not been implemented for most standard setting processes. I have wanted to check how well panelists understand the definitions and the details of the kernel before allowing them to perform the tasks specified by the kernel, but I do not know of any standard setting processes where such checking is regular practice.

It is clear that more research needs to be done to determine if panelists can perform the tasks required of different standard setting kernels. These research studies are not easy to design and perform, but they are critical to the credibility of the results of standard setting studies.

6

Panelist Selection and Training

6.1 Introduction

The theory of standard setting that is the basis for this book is that standard setting is the translation of policy definition, $\mathbf{D}_P$, of a standard to descriptions of knowledge, skills, and abilities for persons who meet the standard, $\mathbf{D}_C$, and then to the language of the reporting test score. The process of standard setting is designed to facilitate this translation and gather evidence to support the credibility of the results. Some implications of the theory are that qualified persons do the translations, and they use various aids for the process, usually called standard setting procedures, to produce accurate translations. Therefore, selecting translators (usually called panelists) and making sure that they have the tools necessary to produce a credible translation is an important part of the standard setting process.

I learned early in my career that selecting a translator that gives a useful and credible translation is not as easy as it would seem. I needed a translation of a technical article on computer applications to educational testing that was written in German. I contacted a faculty member who was in the German department at my university and hired her to do the translation. She was a native speaker of German and was highly proficient in English. The translation she produced was almost unusable because she did not know the statistical and psychometric content and often selected inappropriate words and phrases in English to correspond to technical terms used by the German researcher. The literature on panelist selection for standard setting mirrors my experience with translation from one language to another. That literature shows that the characteristics of panelists have an influence on the level, consistency, and overall credibility of the standards (e.g., Clauser, Swanson, and Harik (2002); Margolis, Mee, Clauser, Winward and Clauser, J. C. (2016); Peabody and Wind (2019); Pitoniak and Cizek (2016); Plake (1998); Plake, Melican and Mills (1991); Raymond and Reid (2001)).

The standard setting literature describes two ways of addressing the problem of how to select panelists. One way is to identify individuals with expertise on the substance of the construct implied by the policy definition of the standard, $\mathbf{D}_P$, and with the cognitive skills for implementing the

DOI: 10.1201/9780429156410-6

requirements of the standard setting kernel (Raymond and Reid (2001)). The second way is to consider identifying panelists as a sampling problem. The agency specifies a domain of appropriate individuals, and the goal is to obtain a representative sample from that domain. An emphasis with this approach is ensuring that the process is replicable. The domain definition often includes characteristics that are not directly related to expertise such as gender, race/ethnicity, region of the country, attitudes toward the standards, etc. Inclusion of these characteristics in the domain definition is believed to enhance the credibility of the resulting standard (Jaeger, 1989). The emphasis on diversity and representativeness is likely to result in more variation in translations and the standard setting process will need to have procedures for keeping the variation to acceptable levels and/or adjust for it.

The domain sampling approach to selecting panelists requires that several statistical issues be considered. How many persons must be sampled for the sample to be considered representative? What sampling design should be used? What sample size is needed to get stable/reproducible estimates of the standard? These statistical issues are different depending on the standard setting kernel used in the standard setting process and the uses to be made of the standard. Higher stakes require greater accuracy/stability of the results than situations with lower stakes. The acceptable level of variation in the translations becomes an important aspect of the design of the standard setting process.

The first part of this chapter summarizes the literature on standard setting related to the selection of panelists and relates that literature to the theory of standard setting proposed here. The second part of the chapter acknowledges that panelists are unlikely to have all the knowledge, skills, and abilities needed to produce a precise translation of $\mathbf{D}_C$ to the reporting score scale. As a result, standard setting processes devote significant amounts of time to training panelists about $\mathbf{D}_P$, $\mathbf{D}_C$, and the details of the kernel used to support the translation process. The second part of the chapter summarizes the literature on the training provided to panelists to facilitate the standard setting process. The content and design of the training are related back to the theory of standard setting.

6.2 The Influence of Panelists' Characteristics on the Estimated Standard

It may seem obvious that panelists must have expertise on the construct implied by the $\mathbf{D}_P$ for them to be able to produce a credible translation. When looking for persons to translate my article written in German into English, I did not consider persons who were not proficient in German. But selecting a group of experts on the knowledge, skills, and abilities implied by the construct underlying the reporting score scale can still result in variation in expertise. Just as my German expert was not familiar with the technical concepts and vocabulary of statistics, experts on mathematics, reading, or medical specialties do not have identical skills and knowledge bases. There is variation in domain expertise of even highly regarded individuals.

It is difficult to accurately determine the level of expertise of panelists. They are usually recruited based on credentials and recommendations from relevant organizations. The research investigating the influence of the knowledge, skills, and abilities related to the construct of interest takes advantage of part of the typical training program for panelists. Panelists are often asked to take either the full test or a short form of the test administered to examinees. This is done to familiarize panelists with the types of tasks on the test and the conditions under which examinees take the test. Often the panelists are given the scoring guides for the test and told to score the test on their own. The results from the scored tests can be collected and used to get a fine-grained appraisal of the panelists expertise on the construct of interest.

The general result from studies using panelist performance on the tests they are using in the standard setting process as the measure of expertise is that standards set by higher scoring panelists are higher than those for lower scoring panelists (Clauser, Swanson and Harik, 2002; Margolis, Mee, Clauser, Winward and Clauser, 2016; Peabody and Wind, 2019). This is even true at the item level for the Angoff kernel. Those panelists who answer an item correctly tend to estimate the conditional probability correct for examinees described by $\mathbf{minD_C}$ as higher than those who answer the item incorrectly (Margolis, Mee, Clauser, Winward and Clauser, 2016). These studies do not address the issue of whether the higher standard is a better representation of $\mathbf{minD_C}$ than the lower standard.

Some insight about the influence of panelists expertise can be gained from the analogy of translation to another language. If a children's book is being translated from one language to another, would the results better reflect the author's intent if the translation were done by a university professor of literature or by a reading teacher who regularly works with students who read a book of that type? Under the assumption that the teacher would be the better choice, many standard setting processes include a requirement that panelists have direct experience with the target examinee population for the test that operationally defines the construct (Kane, 1994; Raymond and Reid, 2001). Panelists' knowledge of the characteristics of the examinee population is expected to give them a more nuanced understanding of $\mathbf{D_P}$ and $\mathbf{D_C}$ than those without that knowledge so they can better translate those definitions to the language of the test score scale. While it seems reasonable to require that panelists have experience with the examinee population, no formal studies have been located that investigate the impact of knowledge of the examinee population on the standard setting process.

The literature on standard setting suggests that in addition to level of competence, panelists should be selected to represent various demographic groups. For example, panelists might represent specific interest and stakeholder groups. Or, it may be important that the panelists match the examinee population in terms of race/ethnicity and geographic region. Including demographic selection criteria for panelists is consistent with a domain sampling conception of selecting panelists. From that perspective, the goal is to get a representative sample of appropriate potential participants. There is some research that indicates that groups that vary on demographic characteristics set standards at different points on the reporting score scale (Linn, Madaus and Pedulla, 1982; Livingston & Zieky, 1989). However, those results are difficult to interpret because there may be confounding between group membership and other characteristics like expertise and experience with the examinee population. NAEP standard setting studies did not show differences between subgroups when the panelists were carefully selected to have similar expertise and experience with the examinee population (Shepard, Glaser, Linn & Bohrnstedt, 1993).

To give a sense of how expertise enters the translation process, consider the following specific example. The following statement is part of $\mathbf{D}_C$ for the mathematics portion of NAEP. "Students at this level should understand the process of gathering and organizing data and be able to calculate, evaluate, and communicate results within the domain of statistics and probability" (NAGB, 2019, p. 73). This description relates to the "data analysis, statistics, and probability" part of the NAEP Mathematics test specifications. The grade level for this description has not been stated to make a point. In many cases, the language of $\mathbf{D}_C$ is very general and open-ended. There seems to be a reluctance to indicate what someone at a level of proficiency cannot do, so the upper end of expected performance is not clear.

Now consider a panelist and the level of expertise needed to properly translate this $\mathbf{D}_C$ into a point on the reporting score scale for the test. At the very least, panelists need to know as much or more about data collection, statistics, and probability as the target population for the examination. But, because of the imprecise language of $\mathbf{D}_C$, exactly what the panelist needs to know is not clear. Knowing the grade level, the mathematics curriculum for that grade level, and the kinds of assignments teachers ask students at that grade level to do adds important detail and context to $\mathbf{D}_C$. It is difficult to imagine how a person without the knowledge of the educational system could perform the tasks required by a standard setting kernel at an acceptable level. The description given above comes from the eighth-grade Proficient Achievement Level Description for NAEP Mathematics. A university professor of statistics with no contact with eighth-grade mathematics education would likely produce quite a different translation of this description than an eighth-grade teacher of mathematics who works with students at that level every day.

The standard setting kernel aids the panelists as they do the translation process. If the Body of Work kernel were used, the panelist would be classifying examples of work as representing capabilities above or below the standard implied by $\mathbf{D}_C$ and $\mathbf{minD}_C$. How would a professional statistician without an understanding of the context for students studying eighth-grade mathematics decide what work is above or below the standard implied by the statement given above? I can imagine giving projects to doctoral students in our Measurement and Quantitative Methods program that would fit with the description coming from the NAEP Achievement Level Description.

The implication of this discussion is that the standard setting process depends on selecting good translators. Those translators must not only have expertise in the subject matter of the test, but also have a thorough understanding of the context for the performance of the target examinee population.

6.3 Selecting and Recruiting Panelists

Given the discussion in Section 6.2, it is important to specify in advance the desired characteristics for panelists. At the very least, panelists need to have the knowledge, skills, and abilities required to understand $\mathbf{D}_C$ and $\mathbf{minD}_C$. They should also know the context for $\mathbf{D}_C$ so that they can elaborate the meaning implied by what are often vague statements of the requirements for the implied standard. It is also important to have the capability of performing the tasks required by the standard setting kernel. For example, persons with no or limited experience with the products of student assignments might have difficulty with the requirements of the Body of Work kernel. Panelists with no experience with probability may have difficulty with the Angoff kernel. The desired characteristics of the panelists should be guided by an analysis of the required tasks for the selected standard setting kernel.

With the descriptions of the desired characteristics of the panelists specified, the task becomes one of identifying individuals with those characteristics and recruiting them to participate in the standard setting process. In some cases, there are readily available lists of potential panelists. For professional certification and licensure, there are lists of those who have passed the examinations in the past and who have relevant experience within the profession. Such lists are a good starting place for identifying panelists. Similarly, when setting standards on state achievement tests, an initial source for panelists are lists of teachers who actively teach in the area of the test. However, not all persons on such lists may have the capabilities to perform the tasks required of the standard setting kernel. Teachers of writing may be excellent at performing tasks required by the Body of Work kernel but have problems with adequately performing tasks required by the Bookmark kernel. Either the kernel should be selected to match the characteristics of the panelists or the panelists selected who can perform the tasks required by the kernel.

It would seem desirable to determine if panelists understand the tasks they are to perform for a standard setting kernel. It is common in the training of persons to score essays, or other open-ended assessment tasks, to check the scores produced by these individuals by comparing the scores to those produced by experts. If the scores do not match those from the experts, the prospective scorers are retrained or excused from the scoring process. I have not found documentation for standard setting processes of evidence beyond self-reports that panelists understand the tasks they are to do or the feedback they receive.

The literature on standard setting describes competing goals when selecting panelists for the standard setting process. One approach is to find individuals who have the necessary knowledge, skills, and abilities. This is consistent with selecting persons for language translation tasks. For language translation, a single person does the translation, and that person is selected because they have high levels of competency in the initial language, the target language, and the genre of the written text. As discussed in earlier chapters, it is difficult to find such persons in the standard setting context because few persons are highly proficient in the language of policy and content standards, and in the language of the reporting test score. To adequately accomplish the translation in the case of standard setting, it is best to think of the selection of panelists as selecting team members that support each other in the translation process—members who are subject matter experts, members who are knowledgeable about the characteristics of the examinee population, members who are experts in the test design and meaning of the reporting score scale, and members who are proficient with the tasks required by the standard setting kernel. There is also a need to replicate the process so that it is not the idiosyncratic result of the panel selection process. Such a panelist-selection process is not a random one, but a careful selection to identify a group that will best be able to accomplish the translation task with the kernel selected to support the process.

The other approach to panelist selection is more consistent with the idea that standard setting is a matter of opinion rather than one of translation. The goal is to get a representative sample of appropriate panelists so that the consensus opinion can be determined. In this case, appropriate usually means representatives of all groups who have an interest in the results of the standard setting. In the educational context, this might include parents and teachers' union representatives as well as curriculum developers and textbook authors. There would also be a need to have gender and ethnic balance because persons from different backgrounds might have different opinions. Some of these selection criteria may not help with a translation process. This becomes more apparent when certification and licensure standards are considered. If the standard is for the certification of automotive mechanics, it does not make sense to have persons on the panel who are automobile owners with little technical background even though they have strong interest in the results of the standard setting. Having the required expertise should take priority over interests and influence.

Some panelist-selection processes attempt to merge these two approaches. For example, the process for setting standards on NAEP first contacted all interested groups asking for nominations of persons with the required characteristics (Raymond and Reid, 2001). Then the credentials of those persons were reviewed to identify a pool of potential panelists. Then multiple samples were drawn from the pool that would meet the expertise requirements plus other demographics that support the requirements for representation of interested groups. This selection process is very good, but it also takes time and planning. Unfortunately, many panelist-selection processes sample from convenient groups under an assumption that all persons who agree to participate have the needed expertise.

6.4 Statistical Considerations for Selecting Panelists

Because it is not possible to find a single translator who has the necessary expertise to translate from descriptions of policy and content to a point on the reporting score scale, groups of persons are selected with the expectation that they collectively will have the necessary expertise, and also under the belief that an "average" of the translations by several persons will be a better estimate of the intended standard than the results from one person. The use of several individuals on a panel may balance out errors in interpretation of policy and content descriptions and in judgments about items, persons, or work samples. However, this approach leads to questions about how many panelists should be selected and what their characteristics should be.

One approach to deciding on the number of panelists is to determine the amount of variation in the standards estimated from the data from individual panelists and then compute the variation in the "average" standard from the group. Some sources label this as the standard error of the standard (e.g., Kane (1994) and Hambleton (1998)), and they suggest that the number of panelists should be sufficient to result in a standard error that is small enough to show that the estimated standard is stable. Also, if multiple standards are estimated on the reporting score scale, the standard error bands around the estimated standards should not overlap. The estimate of the variation for the average that is commonly used, the mean or median, is dependent on the size of the sample that is used to compute it—in this case the number of panelists. If a judgment can be made about the acceptable level of accuracy for the estimate of the average standard, the number of panelists needed to reach that level of accuracy can be computed using the standard formulas for the standard error of the mean or median.

Clauser, Margolis, and Clauser (2014) suggest recruiting 30 panelists for high-stakes standards and dividing them into three groups of 10 to facilitate estimation of the standard error of the estimated standard. However, the number of panelists used in practice varies widely. Boursicot, Roberts, and Pell (2006) used ten panelists for setting standards for medical students. Berk (1986) reports that most standard setting studies use 5–20 panelists.

To give a sense of how to decide on the number of panelists, a simple but realistic example is provided here. Information was collected from several state standard setting studies that used the Bookmark kernel to obtain estimates of translations of the $\mathbf{D_P}$s for Basic, Proficient and Advanced to the IRT-based θ-scales for their tests. The medians of those estimated standards were computed, and they were used as the intended standards for a simulation. The standard deviations of panelists' individual estimated standards from the studies were also collected and the median values were computed. The median standard deviation was close to .2, so it was rounded to that value for this simulation. Table 6.1 gives the θ-values that approximate the median standards from the studies.

To show what these descriptive statistics indicate for a standard setting study, the estimated standards on the θ-scale for 15 panelists using the Bookmark kernel were simulated by sampling from normal distributions with means equal to the intended standards in Table 6.1 and standard deviations of their estimates of .2. The estimates of panelists' standards from one level to the next were assumed correlated .5. This amount of correlation between standards set at different levels has no empirical support. It seems likely that the standards set at one level and standards set at another level by the same panelists will be related. That is partially because they must be in increasing order. The value of .5 was selected as a reasonable amount of correlation, but no information could be found in the literature on the Bookmark kernel that described the strength of relationship between standards set at different levels by the same panelists.

Figure 6.1 presents a histogram that shows the results for the simulated standard setting. The histogram shows that the spread in the bookmark placements of the panelists for the Basic and Proficient standards resulted in substantial overlap in the distributions of estimated standards. Because of sampling error, the estimated standards are not expected to be exactly the same as the values given in Table 6.1. The observed mean and median of the Basic standard were –.66 and –.67, respectively. The mean and median for the Proficient standard were –.28 and –.38, respectively. Similarly, the mean and median for the Advanced standard were 1.24 and 1.23, respectively. The standard deviations for the Basic and Proficient Bookmark placements were .33 and .23, respectively. The standard deviation of the differences between the Proficient and Basic standards over panelists was .22.

TABLE 6.1

Intended Standards for Basic, Proficient, and Advanced on the θ-Scale

	Intended Standards		
	Basic	**Proficient**	**Advanced**
Point on θ-scale	–.65	–.35	1.00

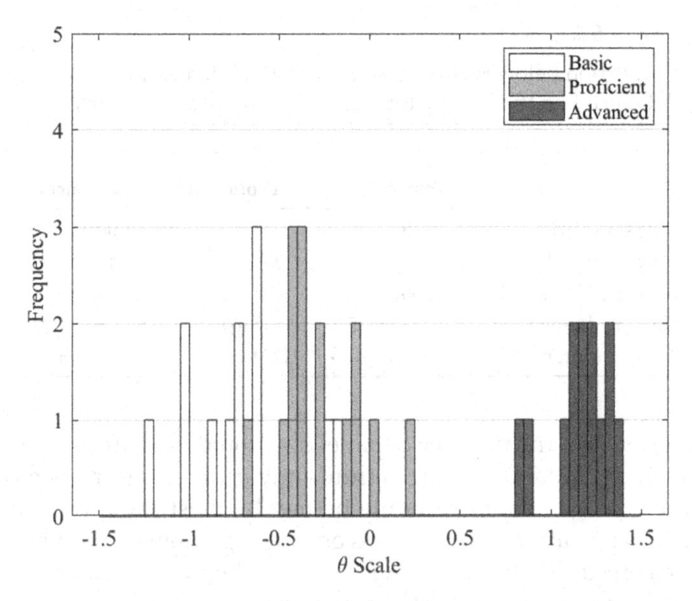

FIGURE 6.1
Distribution of panelists' bookmark placements for three levels of standards.

The information from this simulated sample of panelists can be used to determine how many panelists are needed to get good estimates of the intended standard assuming similar panelists and a similar Ordered Item Booklet (OIB). A minimal requirement is that the estimated standards be significantly different from each other. To check if this is the case, a paired *t*-test can be run between the mean estimates of the standards. For this example, the significant difference requirement for estimates of the Basic and Proficient was easily met. The *t*-statistic for the difference between the means of the panelist's estimates was 6.93, well above the value needed to be significant at the .01 level.

In most cases, the levels of estimated standards will be significantly different from each other even with a relatively small number of panelists, such as the 15 used in this example. But there are also cases when many levels of standards are estimated on a reporting score scale and meeting the minimal requirement for significantly different estimated standards might be a challenge. For example, the Common European Framework of Reference for Languages (CEFR) defines as many as 11 levels of language competence (Council of Europe, 2009, p. 135). These range from minimal competence with a language at the tourist guidebook level to the competence of an educated native speaker of the language. At its greatest level of refinement in language descriptions, the CEFR has 11 levels requiring 10 cut scores. With that number of standards to be set, some of them are likely to be close to each other, requiring a large number of panelists to meet even the minimal requirement of estimates of the cut scores being significantly different from each other.

TABLE 6.2

Proportion below Standards for Basic, Proficient, and Advanced on the θ-Scale for True and Estimated Standards

	Levels		
	Basic	**Proficient**	**Advanced**
True standards	–.65	–.35	1.00
Proportion below	.26	.36	.84
Estimated standard	–.66	–.27	1.24
Estimated range	–.75 to –.57	–.34 to –.22	1.18 to 1.30
Range for proportions	.22 to .28	.37 to .41	.88 to .90

The criterion that the number of panelists should be sufficient to conclude that the estimated standards are significantly different from each other is a minimal one and it should certainly be met by most standard setting processes. It does not apply when there is only a single standard. In many cases, the reported result is the percentage above or below a standard. The accuracy of this type of reported information is sensitive to the estimated location of the standard. Consider the example given above with true standards provided in Table 6.1. Assuming that the true distribution of proficiency is normally distributed with mean 0 and standard deviation 1, the proportion below each standard can be computed. Table 6.2 contains the computed values. The standards that are available for use are those estimated from the panelists' bookmark placements. Those are also given in Table 6.2 from the simulation of the standard setting process. Using the standard error of the mean, the range of estimated standards was also computed. Finally, the table provides the ranges of the proportion below the estimated standards based on the estimated range assuming standard normal distribution of examinee performance.

Most notable from Table 6.2 is that the one standard error of the mean range for the Proficient and Advanced levels does not include the intended value. The case for the Advanced cut score is most dramatic. If presenting the proportion above or below a standard is important, then the accuracy of those values should be used to determine the number of panelists that are needed for the standard setting process. If, for example, it was desirable to estimate the proportion below Advanced to within .01, that is, from .83 to .85, then the corresponding range of estimated standards would need to be from .95 to 1.03. An important question is how much confidence is needed that the estimated proportion below the standard is between .83 and .85. If it is 95% confidence, then the range from .95 to 1.03 must represent approximately two standard errors of the mean around the estimated value. Using the formula $SE = \hat{s}/\sqrt{n}$, where SE is the standard error of the mean, $\hat{s}$ is the estimated standard deviation of panelists' estimated standards, and n is the number of panelists, if the desired SE is .02 and $\hat{s}$ is .25, then the required number of panelists is approximately 156. Even if the level of confidence is only 65%, the number of panelists needed to get that level of confidence is 39.

Another way to look at the issue of the number of panelists needed is from the perspective of a person taking the test to receive an important credential, such as when test results are used for making certification and licensure decisions. In that case, the policy makers are typically trying to determine who is competent to perform in the occupational areas that they manage. $\mathbf{D_P}$ needs to consider the consequences of false positives and false negatives. False positives are letting persons without the expected level of competence enter the profession. False negatives keep persons who are qualified from working in the profession. Well-crafted $\mathbf{D_P}$ should take these issues into account when defining the intended standard.

It is difficult to make general rules about the number of panelists needed for a standard setting study. The number will depend on how the standards are used and the amount of variation in panelists' estimated standards. In the example given above, if the variation in estimated standards was lower, the number of panelists needed would be fewer. That relationship implies that it is important to recruit highly qualified persons to be panelists and that they should have a thorough understanding of the task that they are asked to do. That leads directly to the next topic in this chapter—the training of the panelists for the standard setting process.

6.5 Training of Panelists

Training of panelists is critical to the accurate estimation of the location of the standard on the reporting score scale for the test used to represent the construct implied by the $\mathbf{D_P}$. The panelists have a very challenging task—to translate from the language of policy to the language of the reporting score scale when they most likely have not had a lot of experience doing such tasks. There are professional translators from one language to another who work for governments, businesses, and the United Nations, but there are no professional translators for the task of translating policy to score scales. Professional language translators have years of education and experience, but for the standard setting translation task, panelists may have extensive experience with understanding part of the task, but limited experience with other parts of the task. The purpose of training is to improve the capabilities of the panelists for the tasks they are asked to perform.

According to the theory of standard setting presented in this book, panelists need to have an understanding of, and proficiency with, a number of concepts and tasks. The first of these is to understand the thing that is to be translated—the $\mathbf{D_P}$ and the other information used to elaborate on it, such as the $\mathbf{D_C}$ and $\mathbf{minD_C}$. Without a deep understanding of this information, the panelist's task becomes undefined, likely leading to wide variation in the estimated locations of the standard. Consider the example of the standard for "tall" used earlier in this book. If panelists are given no information about how the standard will be used or the reason for producing the standard, each person must decide on his or her own what $\mathbf{D_P}$ is. A person might simply decide that tall means more height than my own. Another might decide it means someone like a professional basketball player. To have an interpretable meaning for the standard, and a validity argument to support the value on the score scale, panelists need to understand the intentions of the policy making body—in this theory referred to as the agency.

A second concept that panelists need to understand is the meaning of the reporting score scale that is the target language for the translation. It is impossible to translate something into another language unless the translator knows the other language. This is a particularly important part of training because few persons outside of test development organizations know the details about how tests are produced and scored. There are also misconceptions that must be countered so that they do not influence the translation process. For example, a panelist might believe that if a person correctly answers 60% of the items on a test, that they correctly answered the easiest 60%. Such beliefs have serious consequences for panelists' use of some standard setting kernels such as the Bookmark kernel. Some panelists may expect that all items before the placement of the bookmark are answered correctly, and the page number is the equivalent of the number-correct score for the standard. This is not the case unless the test items are highly intercorrelated with each other and are highly discriminating—that is, they form a Guttman scale (Guttman, 1950). Tests used with the standard setting process rarely have these properties.

A common misconception is that the score points on a scoring rubric for an open-ended task correspond to the multiple levels of a standard setting process. For example, a panelist might consider a four-point rubric to directly correspond to Below Basic, Basic, Proficient, and Advanced for a state assessment. This is seldom the case, and such interpretations ignore the differences in difficulty of the open-ended tasks. There may also be misconceptions about the relation between number-correct scores and the reporting score scale that is often based on an IRT model and/or an equating procedure.

A third component that panelists need to understand is the standard setting process itself. They need to understand the activities that have been selected to help them with the translation process. These include the details of the standard setting kernel, the types of information that are provided to them during the process, and the overall structure of the standard setting process. They also need to understand that they are doing a translation task, not setting policy. The panelists need to understand the tasks that they are to accomplish with the standard setting kernel. They also need to understand the meaning and usefulness of information such as feedback and normative data that are provided during the process. I personally have observed panelists who do not understand some of the information that is provided and, because of lack of understanding, ignore that information.

It is also useful for them to understand why they are performing the tasks indicated by the standard setting kernel several times, and how they should interpret differences between their translation and those of others on the panel. Much of the information that they receive and many of the tasks they are asked to perform are not familiar to them, and it is important that they have a level of competence performing the tasks before being asked to give formal recommendations for the translation.

The three main components of a standard setting study—policy, test, and process—are each very complex, and it is unlikely that panelists can learn what they need to know in a short period of time. Most standard setting processes take multiple days to complete, although there is always pressure to make the process as short in length as possible. Using a very short time for the process is a false economy because the resulting estimates of points on the reporting score scale very likely will not correspond to the intended standard, and the estimates may be highly variable.

Several authors have provided detailed guides to the organization of a standard setting process (Cizek & Bunch, 2007; Hambleton, 1998; Pitoniak & Cizek, 2016) and the technical reports from standard setting processes often contain descriptions of activities for each day of a standard setting meeting (e.g., Florida Department of Education, 2020, Chapter 5). These are very useful as guides to designing a standard setting process, but they should not be followed too mechanically. The most important outcome of panelist training is a high level of confidence that they understand all the information provided to them and can competently perform the tasks that are included in the process.

The unsung heroes of the standard setting process are the persons who provide the training and guide the rest of the process. These individuals are usually given the title of "facilitator," because they are to facilitate the successful completion of the process. In some cases, a standard setting process may have multiple facilitators, although they might not all be given that title. In addition to the person doing most of the training and managing the flow of the process, there may also be a content expert and a testing expert. These persons are included to answer detailed questions about the $\mathbf{D}_C$s, and the design and construction of the test that operationally defines the construct intended by the policy definition. As is true for the panelists, it is rare that a facilitator is knowledgeable about all aspects of the process.

As with other training/instructional activities, the capabilities of the persons doing the training make a difference. They must understand the goals for the standard setting session and how all the tasks fit together to achieve the goals. They need to know what is important, how panelists may get off track, and how to manage the entire process. Cizek and Bunch (2007) recommend that the full process be given a field trial to identify potential problem areas and to make sure that the facilitators know how to handle specific situations like panelists ignoring the $\mathbf{D_P}$ or being confused by the tasks required by the standard setting kernel.

The standard setting process is a very complex activity with many requirements. Unfortunately, it is possible to have a poor design for the process and have it implemented by inexperienced or untrained facilitators. Even in those cases, there is the expectation that there will be a usable product at the end. Over the years, I have taken a number of professional development courses, and the closest one to a standard setting meeting was training to use and develop software for expert systems. It was a week-long course, and there was the expectation upon completion participants could use software to model expert systems. Some of the "trainers" were very good. Others were not. At the end, some participants could use the software and others could not. For standard setting, there is not much tolerance for a poor process.

The remainder of this chapter presents information about specific training components for a standard setting process. Much of the design of such processes is more art than science. Learning to design and facilitate such processes is typically acquired through a process much like an apprenticeship with a furniture maker. "Master" facilitators pass along wisdom acquired through working on many standard setting projects to the new facilitators.

6.5.1 Presenting the "Big Picture"

Imagine being recruited to be a panelist for a standard setting study and you have agreed to participate. It probably takes place in a nice hotel in an interesting location. You may see old friends and meet other interesting people. But what were you told to entice you to participate and what do you expect to happen at the sessions? It is important that panelists know why they were recruited and what they will be expected to do. Participating in standard setting is hard work. It is equivalent to translating the Preamble of the US Constitution into another language where it will be used for some very important communications. The use is so important that it is worth carefully selecting every word. The translation must keep the meaning of the document as close as possible to the original.

Initial communications with panelists are especially important. These communications need to set an appropriate tone and lead to realistic expectations for what will occur in the standard setting meeting and how the end results will be used. Cizek and Bunch (2007) recommend that the initial session of the standard setting process include a presentation by a representative of the policy board that called for the standard (the agency). That person should inform the panelists about the importance of the work and how the results of the process will be used. Cizek and Bunch (2007) also suggest that panelists receive information prior to the opening session of the standard setting process, possibly weeks prior to that session, that indicates the goals for the meeting, the agenda showing all the activities, and $\mathbf{D_P}$ and $\mathbf{D_C}$. Some of the NAEP standard setting processes (e.g., Loomis, Hanick, Bay & Crouse,

2000) sent a booklet to panelists prior to the in-person session that included all this information plus a glossary of terms that would be used during the standard setting process.

From the perspective of the theory of standard setting presented here, this initial information is very important. To produce a coherent translation of the $\mathbf{D_P}$ to the reporting score scale, the panelists need to have an in-depth and finely nuanced understanding of the $\mathbf{D_P}$ and the $\mathbf{D_C}$ derived from it. Without that understanding, it is unlikely that the resulting translation will accurately reflect the intentions of the agency.

6.5.2 Training on Policy and Content Definitions ($\mathbf{D_P}$ and $\mathbf{D_C}$)

Even if the $\mathbf{D_P}$, $\mathbf{D_C}$, and $\mathbf{minD_C}$ have been distributed to panelists in advance of the standard setting meeting, it is important to go through these definitions with the panel as a group. The sessions related to the definitions are usually described as coming to a "common understanding" of the meaning of the definitions. This terminology and the process is consistent with the theoretical position that the goal is to produce one translation that accurately reflects the intention of the agency. In some cases (e.g., Mitzel, Lewis, Patz and Green, 2001), the definitions may be elaborated during the standard setting process to help panelists reach the common understanding and improve the match to test content. Producing elaborated definitions is controversial because of the possibility that the results will change the intentions of the agency (Shepard, Glaser, Linn and Bohrnstedt, 1993). However, producing elaborated definitions is an effective way to identify misunderstandings about the definitions. Working on the definitions also helps panelists to gain a "common understanding" of the intentions of the agency. In the ideal case, the agency should approve any elaborations or other changes to the definitions before the revisions are used for the translation process. Sometimes that is not practical, so the definitions provided to panelists are treated as fixed and the focus of training is on having panelists get a common understanding of the approved definitions. In any case, all panelists should have ready access to the definitions during the process because they are the basis for the translation.

6.5.3 Training on the Language of the Reporting Score Scale

Equally as important as training on the policy and content definitions is the training on the meaning of the reporting score scale. The reporting score scale is the target language for the translation and, while panelists may have some experience with scores on the test, they are unlikely to have knowledge and understanding of the nuances of test score-point meaning that are required for an accurate translation of the $\mathbf{D_P}$.

Training on the reporting score scale generally has three components. First, panelists are asked to take the test that operationalizes the construct under the same conditions as the typical examinee. This serves two important functions. It gives the panelists the context for performance of examinees, and it gives them direct experience with the test items so they will know their characteristics. To maximize the usefulness of this activity, the panelists should not have the answer key before taking the test. The test should also be full length even though that may extend the length of the standard setting meeting. Using shortened tests or sample items does not help panelists understand the context of the test for the operational examinees. Nor does giving panelists the answers in advance. It is common to let panelists score their own tests so that they learn about the scoring process.

The second major training component is about the details of the test design. Along with the practical experience of taking the test, panelists also need to know the intended construct that the test score was designed to represent. This means that panelists also need to get training on the test design and development process. Are test items selected to be a representative sample of tasks from a domain of knowledge, skills, and abilities (KSAs)? If so, the panelists should be presented with the domain description and the method used for sampling test items to ensure that the items represent the domain. Panelists should also be given any additional requirements for item selection such as a target range of difficulty or balance in gender representation.

The third aspect of training related to the test is how the examinee responses to test tasks are converted to the reported scores. The test scoring and reporting can be very complex and most casual users of test results don't have an in-depth understanding of how the tests are scored. This is particularly important when test items that generate open-ended responses are used on the test and advanced psychometric procedures such as test equating, adaptive testing, and IRT ability estimation are used for the test. This is a challenging part of training because many of the commonly used scoring procedures can be complex, requiring graduate-level courses in psychometrics for complete understanding. The designers of the standard setting process need to carefully evaluate the scoring process for the test to determine the features that might affect the translation process. Careful training about scoring is also important for understanding some of the feedback that is typically given during the standard setting process. For example, if differences in panelists' estimated standards are provided as feedback, it is important that panelists understand the scale that is used for the feedback and how estimates were obtained. Chapter 7 gives more information about the kinds of feedback given during standard setting.

6.5.4 Training on the Tasks Required of the Standard Setting Kernel

A critical component of panelist training is in the use of the supports provided for the translation process. This component includes training on the tasks specified by the standard setting kernel, the interpretation of feedback provided during the standard setting process, the structure of the entire standard setting process, and goals to be achieved at the end of the process. The goals of the process include a description of the products that will be delivered to the agency after completion of the process.

This component of training often begins with a discussion of the full agenda for the standard setting process and the goals to be achieved. This gives the panelists the big picture related to all the tasks that they will be asked to do. Then it is important that they learn the requirements of the standard setting kernel. This is the basic translation process, and they need to have a reasonable level of proficiency doing the required tasks before starting the translation process itself. Raymond and Reid (2001) note that many panelists will not be familiar with the tasks required by the standard setting kernel. They state: "It is unlikely that participants will understand or correctly perform the required tasks without benefit of training" (p. 140).

Training on the feedback panelists receive about the results of the process is also important because feedback provides a mechanism for refining/correcting panelists' understandings of the requirements of the standard setting kernel. Training related to the use of feedback is often provided immediately before the feedback is presented. Training on feedback is particularly important because practitioners report that panelists ignore or misuse feedback if they do not understand what it means or its usefulness for the translation process (Harrison, 2015).

6.5.5 Planning and Timing

The previous discussion of the training that needs to be conducted to support a credible translation of $\mathbf{D_P}$ to a point on the reporting score scale for a test shows the requirements for the training process. Conducting a standard setting study that yields quality results is a challenging undertaking that needs thorough planning, prior preparation of materials, development of data analysis systems, and careful training of the staff who support the process—facilitators, data analysts, logistics support, etc. It is not the function of this book to provide detailed menus for activities to be included in a standard setting process. It is about the theory behind the process, not the many details that need to be considered when implementing a standard setting process. Useful sources for those details are Cizek and Bunch (2007) and Cizek (2012). Here, the planning of a standard setting session is discussed from the perspective of how the activities support the capacities of panelists to function as translators of $\mathbf{D_P}$ to the language of the reporting score scale.

One of the practical issues related to standard setting is the amount of time available for conducting the process. If an expert bilingual translator were given a document to translate, time would be managed as any other task of that type. The prior experiences of the translator would allow him or her to provide a reasonably accurate estimate of how long it would take to produce the usable translation and they could work back from the date it is needed to set a work schedule. But in the usual standard setting process, the panelists are not expert translators, and they need various kinds of support to help them perform the required task. Also, multiple panelists are used to compensate for the lack of expertise. The expectation is that the collective judgments improve the precision of the translation.

These specific characteristics of the standard setting process make a standard setting session more like a training course for a specific activity than getting a group of people together to get a survey of their opinions. A challenge to planning a standard setting activity is that it is usually not possible to get panelists to dedicate more than a few days to the activity—at most a week. Training courses are usually much longer than that—weeks or months. Designers of standard setting processes must determine how to transmit the required information to panelists in a very efficient way.

One approach is to send information to panelists before the actual standard setting session and hope that they will give it thoughtful consideration. Good candidates for pre-meeting materials are the full agenda with a description of the purpose of each activity, and documents describing $\mathbf{D}_P$ and $\mathbf{D}_C$, and the specifications for the test that defines the construct of interest. There should also be clear expectations that panelists will carefully study these materials and come to the standard setting session prepared to intelligently discuss the materials. There might even be a preparatory task asking them to identify and describe persons who meet the minimum qualifications for the standard. Then they would be ready to work on the common understanding of the $\mathbf{D}_P$ and $\mathbf{D}_C$ before coming to the standard setting session.

I wish I could evaluate each panelist's understanding of these materials before moving on to the actual translation task and dismiss those who did not have a good understanding. Unfortunately, that does not seem to be acceptable and as a result panels tend to be made up of people with mixed levels of understanding of the materials. This tends to result in variation in the estimation of the points on the reporting score scale.

Another important activity prior to the standard setting session is recruiting facilitators and training them in their role. Facilitators are both instructors and managers. They need to be knowledgeable about all aspects of the process and be able to clearly communicate information to panelists. They also need to be good listeners and be able to respond to questions and concerns from panelists. They are also the persons in charge of managing the timing of the process to get it done in the time available. If there are multiple facilitators, they need to have a common understanding of the process. The goal is to avoid what is called a "facilitator effect" (Clauser, Margolis & Clauser, 2014). A facilitator effect is a difference in standards set by separate groups of panelists due to the functioning of the facilitator.

The standard setting process itself is usually organized into a set of related tasks. These should all be organized around the goal of helping panelists produce an accurate translation of $\mathbf{D}_P$ to a point on the reporting score scale. A challenging part of the design of a standard setting process is determining the amount of time to allocate to each task. There needs to be enough time for panelists to develop a deep understanding of each piece of critical information. The parts of the process that tend to take the most time are gaining a common understanding of the $\mathbf{D_P}$ and $\mathbf{D_C}$, the tasks required by the standard setting kernel, and understanding the feedback that is provided to refine the translation. Of course, panelists also need time to perform the tasks required by the standard setting kernel. Given the level of learning required for most of these tasks, the time planned for them should be generous. The tasks require discussion to get the common understanding and practice to make sure panelists understand what they are to do before doing the formal standard setting task. It is difficult to conceive of a standard setting process that would take less than two days. There are examples in the standard setting literature that take less time (e.g., Boursicot, Roberts & Pell, 2006). They give minimal training and seem to assume that panelists already know everything that is required to do the translation. The result is likely a poor translation. It would be like translating the book title *Le tour du monde en quatre-vingts jours* (Around the World in Eighty Days) into "A tour of the world in a number of days." In some sense the translation is accurate, but it does not capture the main theme of the title.

Training for the use of feedback deserves special mention because of the importance of feedback for improving the accuracy of the estimated standard and the challenges panelists face in understanding feedback. The next chapter discusses feedback in detail. Here it is sufficient to indicate that feedback is generally of two types. The first is feedback about the proper performance of the task required by the standard setting kernel. This feedback is designed to help panelists determine if they are doing tasks required by the standard setting kernel in the way that is expected. The second type of feedback is more correctly labeled as context information. This feedback provides additional information about the tests and/or the examinees that is designed to help panelists better target their translation to the audience and purpose of the standard. For both types of feedback, it is important that panelists have in-depth understanding of the meaning of the feedback, and information about how the feedback can be used to help them refine their translations of $\mathbf{D_P}$ to the reporting score scale. As with the other aspects of training, helping panelists understand and use feedback takes time, and panelists who are not familiar with the feedback require more time to gain the required level of understanding.

Most standard setting processes are structured so that training is given before each new task is presented. There is training about the substance of $\mathbf{D_P}$ and $\mathbf{D_C}$ before having discussions aimed at gaining a common understanding of the definitions among panelists. There is training about the test before asking panelists to review the test itself. There is training about the standard setting kernel before panelists are asked to perform the tasks required by the kernel. So it is with feedback. Usually, there is training in the meaning and use of feedback before the feedback is provided to panelists, and additional training is given before each unique type of feedback. One concern about all this training is that panelists may get lost in the details of each individual task and lose track of the overall goal of the process. This means that part of the training is about the context for the tasks and the way tasks were designed to lead to the final product.

6.5.6 Evaluating Training

Most standard setting processes include a number of evaluation activities during the process. These evaluations are performed because authors such as Kane (2013) have indicated that evidence that the process was well structured and implemented is needed to support the validity argument for the standard. For the most part, this evidence comes from surveys of the panelists to determine if they understood what they were to do and if they had enough time to do it. These surveys typically ask for levels of agreement with statements like "The timing and pace of the method training were appropriate"; "The training in the standard setting method was clear"; and "I understand the kinds of feedback that will be provided to me during the standard setting process" (Cizek, 2012). There are often surveys of panelists' understandings after each major component of the standard setting process. It is important to note that these surveys do not provide evidence that panelists really do understand these parts of the process, but only that they believe that they do. It is an important research question whether panelists who state that they understand the feedback really do understand the feedback. There are few research studies on this topic. Those that do exist tend to be small, qualitative studies that gauge panelists' understandings through focus groups or interviews (e.g., Shepard, Glaser, Linn & Bohrnstedt, 1993; Skaggs, Hein & Wilkins, 2020). The results from those few studies are not very optimistic. They indicated that panelists found the tasks required of the standard setting kernel to be difficult and many developed simplified methods to accomplish the required task. Moreover, the task that panelists accomplish may not be the task that is desired. For example, rather than translating the $\mathbf{D}_P$ and $\mathbf{D}_C$ to a point on the reporting score scale, panelists may implement a rule such as "70% correct for a test is passing" instead of doing the actual translation. Or panelists may ignore feedback because they are not sure how to interpret the information.

The evaluative surveys do provide useful evidence related to the validity argument for the standard because they indicate if panelists had sufficient time to perform the required tasks and if the process was well run. This is important information. However, these surveys do not give convincing evidence that panelists understand the tasks before them and can properly do them. It is interesting that for almost all standard setting studies panelists will indicate that they understood the information provided to them and the tasks they were to do. They will also indicate that the process was well organized and implemented. There must be a very serious problem with the standard setting study before panelists' ratings move away from the positive ratings.

6.6 Conclusions

There is no doubt that performing the tasks required for translating $\mathbf{D}_P$ and $\mathbf{D}_C$ to a point on a recording score scale is challenging. That high level of challenge does not mean that it cannot be done. Any translation from one language to another is challenging. Most individuals do not have the knowledge and skills to perform translations. But there are many people who do translation from one language to another as their full-time job. I am sure they are not equally good at their job, but many are excellent. The same is true for performing the tasks needed for standard setting. Most persons do not have the knowledge and skills to do the task (although the literature on standard setting suggests randomly sampling persons for the task), but some persons do. It is the goal of recruiting panelists to identify persons with the necessary knowledge and skills who can also absorb the information from training in the amount of time dedicated to the training. Selecting those persons who happen to be available at the scheduled time will not likely result in a panel of people who are competent to do the task.

The research literature on standard setting is beginning to show the importance of panelists' substantive content knowledge on the results of the standard setting process (e.g., Peabody & Wind, 2019). This is not a surprising finding, but it is good to have it documented. There are also many templates and lists of recommendations for how to organize the training for standard setting processes. Reports of the details of well-run standard setting processes can be used as guides for the design of standard setting processes for new situations. The websites of the agencies calling for the standards often have links to reports of their standard setting processes.

From the perspective of the theory of standard setting presented in this book, the major message of this chapter is to consider the standard setting process as one of translation of policy to the language of the reporting score scale. To obtain an accurate and credible translation, qualified people need to be selected to do the task and they need to be given the background and training needed to do the task. This means that the designers of the standard setting process need an in-depth understanding of the requirements of the translation task and the entering capabilities of the panelists. Then, it is important to design a training component that will begin with the entering capabilities and then enhance them so that panelists can competently perform the required tasks. It is very unlikely that a randomly selected group of individuals will have the necessary skills and knowledge to do the required tasks without a quality training program.

7

Feedback and Auxiliary Information

7.1 Introduction

The term "feedback" is often used in conjunction with standard setting processes to describe information presented to panelists to help them understand the results of their judgments related to the standard setting kernel. Feedback seems to have been provided as part of standard setting processes for as long as they have been a formal activity. Nedelsky (1954) suggested pausing after rating five or six items to have a discussion among panelists to determine if everyone has a common understanding of their task and if the results were reasonable. Nedelsky (1954) did not use the term "feedback"; so it is an interesting question of when that term came into common usage. It is now used as a shorthand term to describe any information given to panelists after they have done the tasks required of the standard setting kernel. Designers of standard setting processes have been quite creative in selecting the type of information to provide as feedback and how that information is described to panelists. This creativity is both a blessing and a problem because it is sometimes not clear what purpose the information is to serve in helping produce accurate translation of $\mathbf{D}_P$ to a point on the reporting score scale. Texts on standard setting (e.g., Cizek, 2012) now categorize feedback into several types. Later parts of this chapter describe these types in more detail.

Before discussing the types of information labeled as feedback, it is useful to have a formal definition of the term. According to the *Random House Dictionary of the English Language (2nd Edition Unabridged)* (1987), "feedback" has six different definitions. Many of the definitions are technical and have to do with the way the term is used in the electronics literature. However, the definition from the psychology literature seems most relevant to standard setting. It is "knowledge of results of any behavior, considered as influencing or modifying future performance." In the case of standard setting, behavior is the activity required by the standard setting kernel and the results range from the individual ratings to the estimates of points on the reporting score scale. The definition also indicates that the reason for giving the information is "influencing or modifying future performance." It may seem obvious that

DOI: 10.1201/9780429156410-7

the reason for giving feedback is to influence future activities related to the standard setting kernel, but some designers of standard setting processes worry that the feedback may have too much influence.

This definition of feedback limits the information to "results of any behavior." Typically, this is interpreted as the behavior of panelists as they do the required activities. But some of the information that is provided and labeled as feedback is only remotely related to the behavior of the panelists. For example, Shepard, Glaser, Linn, and Bohrnstedt (1993) suggest looking at the results from other tests when setting standards for NAEP. They mention advanced placement tests, college entrance examinations, and results from international studies of student learning. The intent is to give panelists or policy makers a broader contest for the translation to the reporting score scale. It is now common to provide the estimated proportion of persons above the standard for a variety of subgroups of the examinee population and label that as feedback as well. In a sense, those proportions are a result of the behavior of the panelists, but they are a step removed from the reporting score scale because they are based on a particular distribution of examinee performance.

In this chapter, a distinction will be made between information that is directly related to the performance of tasks required by the standard setting kernel and information that is provided to give a broader context for the translation of the standard. The former is typically designed to help panelists understand the task they are asked to do. The latter are more closely related to understanding the nuances of $\mathbf{D}_P$. It is important to determine how much panelists should be involved in the interpretation of policy. To distinguish between these two types of information, the former (task related) information will continue to be labeled as feedback. It is directly related to the tasks that panelists are to perform. The second type of information (context related) is labeled as auxiliary information. It is information from outside the standard setting process. The auxiliary information is often provided to help panelists understand the $\mathbf{D}_P$ so they can better translate the policy. Auxiliary information does not help panelists understand the tasks required of the standard setting kernel, but the information may help understand the $\mathbf{minD}_C$.

7.1.1 Categorizations of Feedback and Auxiliary Information

The literature on standard setting contains several different labels for the information that is provided to panelists. One set of labels has been borrowed from the literature on the interpretation of test scores: *criterion referenced* versus *norm referenced* interpretations (Cizek, 2012a). In that literature, these terms describe the reference point for test scores used when deciding whether it represents strong or weak performance. Criterion referenced means that the reference point is some criterion for excellence. Many standard setting processes are labeled as criterion referenced because they are translating a specific criterion description to a point on the score scale. Norm-referenced comparisons use the performance of a defined group of individuals to define a reference point for comparison. The group is often labeled as the "norm" or "normative" group and either some measure of average or percentile points from the distribution of scores is used as the basis for comparison.

When these terms are used to describe feedback, the initial meanings of the terms are generalized to the situation of standard setting. Criterion-referenced feedback refers to information provided to panelists about what persons who are near the estimated standard can do related to the content of the test. Norm-referenced feedback provides information about the proportion of a defined norm group that is above or below a standard, or the proportion of the norm group that can give good responses to the tasks on a test.

Another set of labels is used to describe the purpose of giving feedback to panelists. For example, Cizek and Bunch (2007) use the terms "reality feedback" and "impact feedback" along with normative feedback. Panelists receive reality feedback because of the concern that they will have unreasonable expectations about what examinees know and can do. The feedback is intended to connect them back to the "reality" of what examinees actually know and can do. For example, a panelist may look at the items on a test and specify that the minimally qualified person should be able to answer all of them with at least a .90 probability of correct response. Reality feedback might be that the average proportion correct for examinees in the population is .50, much lower than the initial item rating. The expectation is that the panelist would understand that the feedback was indicating that they had too high an estimate of performance and should consider lowering the probability of correct response for some items in the next round of the process.

Impact feedback provides panelists with the proportion above or below the estimated standard so they can evaluate whether the proportion is consistent with the intention expressed by the $\mathbf{D_P}$. There may be concern that giving impact feedback may stimulate panelists to consider other factors besides the $\mathbf{D_P}$. They may think about the political impact on schools, or the need for certified professionals in their field of expertise. Unless the agency specifically included those factors in the $\mathbf{D_P}$, giving impact feedback may result in variations in the translations to the reporting score scale related to the beliefs of the panelists. They change from being translators to policy makers.

Sometimes the term "consequences feedback" is used instead of "impact." Using the term "consequences" emphasizes to the panelists that their judgments have consequences and are very important. Cizek and Bunch (2007) emphasize the importance of the labels given to standards and other information in the standard setting process. The words used to describe feedback, and the standards themselves, make a difference in the way the panelists interpret the information that is presented to them.

Finally, a distinction is sometimes made between process feedback and normative feedback. Harrison (2015) states that feedback "ensures that panelists understand the process" designed into the standard setting. In this context, the word "process" means the steps that panelists perform to properly do tasks indicated by the standard setting kernel. The required tasks may be unfamiliar to panelists and feedback can be designed to help them understand what they are supposed to do.

Even though there are many kinds of feedback used in standard setting processes and many different labels for the feedback, there are two overriding principles behind the use of feedback. The first is that panelists receive feedback that makes sense to them. For the feedback to be effective, it is important that panelists clearly understand the provided information. This implies that the way that feedback is described and the amount of time available to digest the feedback is as important as the feedback itself. Panelists tend to ignore feedback that they do not understand. Or, even if they understand it, if there is not sufficient time to thoroughly review the feedback, they may quickly review it without identifying key details that highlight parts of the process that merit further review, or that show inconsistencies in how they have performed the tasks specified in the standard setting kernel.

The second important principle about the provision of feedback is that panelists need to know the actions that they can take to address issues that they have discovered from the feedback. For example, if there is feedback on impact and the panelists decide that they were setting a standard that was too high or too low, they should be given information about how to change their ratings to make the desired adjustments. They may agree that the feedback identified a problem, but they may not know how to address the problem. Part of the training about the information in the feedback should be how to use the information to improve the quality of the translation of $\mathbf{D}_P$ to the point on the reporting score scale.

7.1.2 Purposes for Providing Feedback

Almost all standard setting processes now have multiple rounds of application of the standard setting kernel or of different kernels. Between rounds there is almost always something called feedback. In some cases, the feedback is given a different label, such as impact, consequences, consistency, etc. One thing that is not always clear is the intended purpose of the feedback. Generally, there are two major purposes for feedback. The first is to help panelists understand and improve on the way they perform the tasks related to the standard setting kernel. This is the type of feedback given any time a person is learning how to do a task or is trying to perfect his or her ability to do the task. When I was taking a course in German, the instructor would let me know if my pronunciation was incorrect or if I was properly spelling words and structuring sentences with proper grammar. Good teaching will provide constructive criticism and guide students to produce better work. So it is with the standard setting process. Panelists need training to perform the tasks required of the standard setting kernel and part of the training is providing feedback on their level of understanding and proficiency with the tasks.

The second major kind of feedback is given to broaden the perspective of persons as they perform the tasks required by the standard setting process. The equivalent type of feedback for my example about learning German would be giving reading assignments from various kinds of texts to show different styles of writing that are acceptable or to show how writing style has changed over time.

When designing a standard setting process, it is important to consider the goals for giving feedback. Is it likely that panelists will have some difficulty doing the tasks required by the standard setting kernel? The literature on the use of the Angoff probability estimation kernel suggests that panelists may have difficulty estimating the probability that a minimally qualified examinee will respond correctly to a test question. If that is the case, then it might be desirable to give "feedback" on the capabilities of specific minimally qualified individuals after discussion of $\mathbf{minD_C}$. It might also be useful to practice the probability estimation process on carefully selected test items and then give the empirical difficulty values as feedback to enhance the value of the practice probability estimation task. In the best of all worlds, the panelists would continue to receive additional practice item sets until they were judged competent at the tasks required of the standard setting kernel. Then the panelists could proceed to the formal standard setting process.

The information provided to give a wider perspective on the tasks required by the standard setting process is not feedback in the regular sense of helping with the training of panelists. It is information provided to help panelists have a deeper understanding of the $\mathbf{D_P}$. For example, if the policy definition states that the standard is at a high level, that can mean either that the minimally qualified person can correctly respond to most of the tasks on the test or that such a person performs better than most other candidates. Even if the $\mathbf{D_P}$ has been accurately translated into a $\mathbf{D_C}$ that represents the desired level of performance, it might be useful for panelists to have additional information to help understand the nuances of the intended standard. For example, after estimating the location of the standard on the reporting score scale, useful feedback might be the score distribution for an operational sample of examinees. The panelists could then observe whether the minimally qualified examinee correctly answers most of the tasks on the test or has higher performance than most of the other examinees. If neither is the case, there is an inconsistency in the connections between the definitions, the test, and the ratings provided by panelists. If the panelists have a good understanding of all this information, they should be able to identify the source of the inconsistency and think of ways to reduce the inconsistency.

Designers of standard setting processes are sometimes reluctant to include score distributions or impact data like the proportion of examinees below a standard because of a concern that panelists will fall back on old rules of thumb like "70% correct on a test is passing" instead of actively working to translate D_P to a point on a score scale. Panelists may also bring in other factors such as empathy for the feelings of examinees if impact data show that a sizable proportion of examinees do not meet the standard. Because of these concerns, designers of the standard setting process seek approval from the agency before including presentations of such information in the process and tend to present it later in the process so it will not have undue influence on the translation process.

There are two other issues about feedback that frequently appear in the literature on standard setting. One is whether feedback is prescriptive. The second is whether the goal of a standard setting process is to reach consensus. Being prescriptive means that the message sent to panelists by the feedback they receive is that they should do something to meet some requirement. For example, suppose that a panelist gave an Angoff probability rating to an item of .20 indicating that the item was very difficult, and the minimally qualified examinee is not likely to answer it correctly. Suppose feedback provided after the round of ratings indicated that the guessing level on the item was .30 because many examinees could eliminate some responses to the multiple-choice item as being obviously wrong. Should the panelist feel required to raise the rating on the item? Similarly, a panelist might receive feedback that the point on the reporting score scale estimated from his or her ratings is much higher than the other panelists. Should the panelist feel that he or she should change the ratings to lower the standard?

Cizek and Bunch (2007) recommend that panelists be told that they do not have to change their ratings after getting feedback. The information provided to them is for them to use in any way they choose, or to ignore. They give this advice because of the concern that panelists will believe that someone has predetermined a standard and that the facilitators are trying to get them to arrive at that standard. The examples in the previous paragraph suggest that the panelist receiving the feedback has some misunderstandings. In the first case, the panelist may not have adequately understood the strategies that examinees use when taking multiple-choice tests. That is, examinees who are low in the target construct can still have a probability of responding correctly to a test item on material they do not know because of guessing strategies. For this panelist, the feedback is prescriptive. The panelists should review the item again to get a better understanding of how it functions and how examinees approach it, and then re-estimate the probability given his or her improved understanding of interaction of examinees with the item.

There are two ways of thinking about the second example. If all the panelists have the necessary skills and knowledge and are doing their best to translate the D_P using the standard setting kernel, their translations should be similar. But the results show that one's translation is notably different than the others'. This means that either the panelist in question has a better understanding of D_P than the other panelists or the panelist has misunderstood some part of the D_P or the steps in the standard setting process. It is the purpose of the feedback to inform the panelists of the situation so they can do something about it. In this sense the feedback is prescriptive. The panelists should review the policy definition, the content-specific definitions, and the use of the standard setting kernel to determine why there is a difference in the estimated location of the standard. If the reason for the difference is a misunderstanding, then the panelist feedback should provide guidance on how to change judgments required by the standard setting kernel.

The second way of thinking about feedback relates to whether the standard setting process should try to bring panelists to a consensus. The reason for having multiple panelists is to compensate for the belief that none of the panelists are experts at the translation process and their collective work will give a better translation of $\mathbf{D}_P$ than having one person do the translation. Further, those running the standard setting process will use statistical procedures to estimate a single point on the reporting score scale to use as the standard, so a value that is the consensus of panelists' efforts will be used whether or not panelists consider it as a consensus. Reviewers of standard setting processes for NAEP (Shepard, Glaser, Linn & Bohrnstedt, 1993) argue that panelists should work to come to a consensus. Yet, some researchers state that "the intent was not to have judges reach consensus" (Clauser, Margolis & Clauser, 2014, p. 131). De-emphasizing consensus results from a concern that the dynamics of discussions among panelists may result in biases in the final estimates, perhaps due to some panelists having more influence on the consensus building process than others.

Even though many articles about standard setting procedures recommend against seeking consensus among panelists, many reports of standard setting studies investigate the level of consensus reached by panelists (e.g., Clauser, Clauser & Hambleton, 2014; Mitzel, Lewis, Patz & Green, 2001; Sireci, Hambleton & Pitoniak, 2004). There are many statistics used to judge the amount of consensus and most standard setting reports discuss the improvement of consensus over rounds of the process. The assessment of consensus is discussed in detail later in this chapter. The literature on standard setting clearly shows that standard setters value consensus. The theory of standard setting proposed here is consistent with that view. If panelists are producing quality translations of $\mathbf{D}_P$ to points on the reporting score scale, the expectation is that those estimated points should be similar.

7.2 Types of Feedback

Although there are various types of feedback with a variety of purposes and they may not accurately fit into a small number of categories, it is helpful to organize feedback into two main types with subcategories in each. The types are feedback about the standard setting process and feedback related to the broader context of the standard setting policy. The feedback about the standard setting process is described and discussed first. Two subcategories of process feedback are delimited—feedback about the understanding of the tasks required by the standard setting kernel and feedback about the similarity of panelists' translations of $\mathbf{D_P}$ and $\mathbf{D_C}$ to the reporting score scale.

7.2.1 Feedback about Understanding of the Standard Setting Kernel

The standard setting kernel is designed to give panelists a task that will help them with the process of translating $\mathbf{D_P}$ and $\mathbf{D_C}$ to a point on the reporting score scale. The kernel will only serve this purpose if panelists understand the required tasks and can competently perform them. One of the concerns expressed about standard setting in general is that panelists do not have a deep understanding of the tasks specified by the kernel and as a result the estimated standard does not accurately reflect the intentions of the policy (Shepard, Glaser, Linn & Bohrnstedt, 1993; Skaggs, Hein & Wilkins, 2020). The practical implications of these concerns are that panelists must be trained to perform the tasks specified by the standard setting kernel and they need feedback about their performance, so they know if they understand their task and are giving useful results that lead to an acceptable translation of $\mathbf{D_P}$.

As indicated in Chapter 5, the interaction of panelists with the tasks specified by the standard setting kernel results in data used to estimate the location of the standard on the reporting score scale. The formal expression for the results of the interaction with the kernel is given by

$$\mathbf{R_i} = K(\mathcal{P}_i, \text{Test Specifications}, \mathbf{minD_C}, \mathbf{D_C}, \mathbf{D_P}), \quad (7.1)$$

where $\mathbf{R_i}$ is the data string provided by Panelist i,
K is the kernel used in the standard setting study,
$\mathcal{P}_i$ is the individual panelist,
and the other symbols have their previous definitions.

When considering process feedback, and specifically feedback related to panelists' understanding of the tasks related to the standard setting kernel, the design question is "What kinds of information will help panelists understand the task that they have been given?" The most basic of these is the distribution of the individual elements in $\mathbf{R}$ over panelists. If all panelists are using the same $\mathbf{D_P}$, $\mathbf{D_C}$, and $\mathbf{minD_C}$ and have a good understanding of the task specified by K, they should arrive at similar entries to $\mathbf{R}$. However, the task specified by K is challenging and the definitions given by $\mathbf{D_P}$, $\mathbf{D_C}$, and $\mathbf{minD_C}$ often have some ambiguities and are open to interpretation. Panelists may also have different levels of understanding of their task. The result is that there are variations in the elements of $\mathbf{R}$ over panelists. The panelist with information in $\mathbf{R}$ that is most closely related to the definitions is not known. The expectation is that if panelists receive the distribution of entries in $\mathbf{R}$ and discuss any differences, they will reduce misunderstandings in all aspects of the use of the kernel. For the expected results to occur, the information provided as feedback must be clearly presented along with a good explanation of why it is being provided. There must also be sufficient time for panelists to discuss and reflect on the information.

A practical example may prove helpful at this point. Suppose a standard setting process is using the Bookmark kernel with a mapping probability of .67 corrected for guessing. Panelists are given training on $\mathbf{D_P}$, $\mathbf{D_C}$, and $\mathbf{minD_C}$ and the standard setting process, and then are asked to work through the ordered item booklet (OIB) to identify the first item that has a probability of correct response of less than .67, ignoring the effect of guessing on the multiple-choice items for persons described by $\mathbf{minD_C}$. Then panelists are asked to continue through the OIB to see if there are any other items that they think the minimally qualified person could answer with a probability of greater than .67. They are asked to note the last of these items. In this case, $\mathbf{R_i}$ has two entries for each panelist: the page number of the item that was before the first identified and the page number for the last item estimated to have a probability greater than .67 for a person described by $\mathbf{minD_C}$. The panelist's estimate of the standard is the midpoint between the mapping points on the θ-scale for the two items selected by each panelist.

The simplest feedback to panelists is providing them with the midpoint between the two bookmarks from the OIB selected by the panelists. This feedback shows the approximate location of the standard in terms of location in the OIB. Figure 7.1 provides an example of this type of feedback for a standard setting process on a 60-item mathematics test used for identifying students to receive advanced placement at a university. The test is a fairly difficult one with an average number of correct responses of 26.82 out of the 60 items. The panelists had been given the university's advanced placement policy statement and a description of the knowledge and skills students were expected to have after finishing the first mathematics course in a sequence. Figure 7.1 shows that the median midpoint bookmark placement was on page 28 in the OIB, but there were panelists whose midpoints of their bookmark placements were as early in the OIB as page 20 and as late as page 40.

Most standard setting processes tell panelists where his or her information is located in the feedback, but do not tell the identity of the panelists providing the other locations. The panelist who placed the bookmark on page 40 knows this is his or her location, but the other panelists do not know who placed the bookmark there unless that person volunteers the information. This seems to be done so as not to embarrass that person, or to reduce the pressure on them to change the bookmark location.

The panelists' likely have many different reasons for their selection of OIB pages. They may have different understandings of the tasks required by the Bookmark kernel and the meaning of the .67 mapping probability; they may have different understandings of the meaning of $\mathbf{minD_C}$; they may have differences in the understanding of the content of the test items and the way examinees approach the items. There are certainly many other explanations.

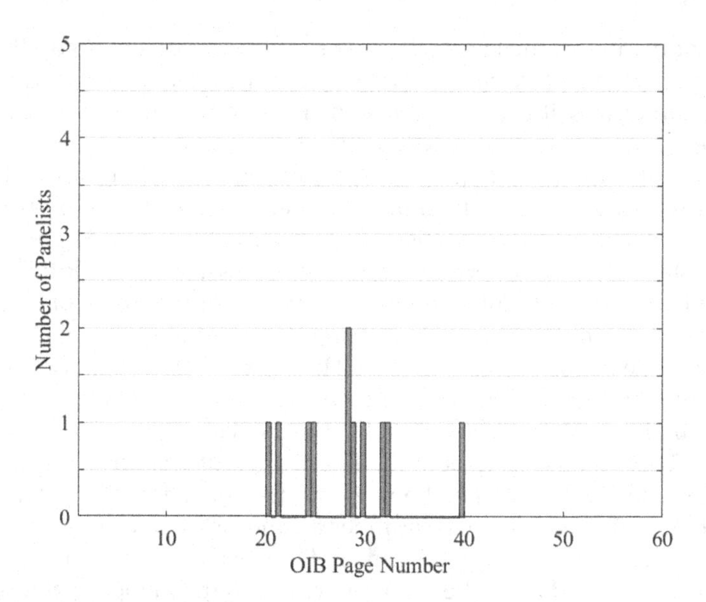

FIGURE 7.1
Panelists' first round midpoint bookmark placements on the mathematics test.

The important question is how will providing feedback about bookmark locations help panelists improve their understanding of their translation task? There can be a discussion after panelists receive the listing of page numbers so they can explain why they made their choice. But, if there are differences in the explanations, how will they decide if they need to change the way they are thinking about things? Is the explanation of any one panelist as good as another panelist?

Because of the many possible explanations for the differences in panelist bookmark placements, there are different answers to the above questions depending on what can be discerned about the panelists' understandings from the data that are available. For example, if a panelist states that he or she placed the bookmark because the task was to put it 67% of the way from the front of the OIB (on page 40 of 60), or the place where 67% of examinees would score below that point on the test, these are inaccurate understandings of the task specified by the Bookmark kernel. In such a case, it is important that the facilitator note this misunderstanding and work with panelists to help them understand the process specified by the kernel.

If the panelists have a different interpretation of the meaning of $\mathbf{D}_P$, $\mathbf{D}_C$, or $\mathbf{minD}_C$, it would be helpful for representatives of persons who wrote those documents to be available to clarify the intent of the definitions. Those representatives should be the authority on the meaning of the definitions, not the views of specific panelists. To make such clarifications possible, standard setting sessions often include representatives of the agency or content experts in the sessions for the purpose of answering questions about the nuances of the intentions behind the information in the definitions. The role for those individuals is to help the panelists get a common understanding of the intentions of the agency.

In some cases, panelists place a bookmark on an item that has some flaws that make it seem more difficult than it was for the students. In the example given in Figure 7.1, the panelists with the lowest initial bookmark placements put their first bookmark on page 10 in the OIB. This item had a low item response theory (IRT) discrimination parameter estimate indicating that it did not clearly distinguish students who had the necessary skills and knowledge from those who did not. However, most students answered this test item correctly. They were able to work through the ambiguous information in the item or possibly there was only one reasonable answer choice even though the test item was about a difficult concept. The expectation is that the group discussion of those bookmark placements will clarify how students approach such items and help panelists improve the quality of information they provide for estimating the standard.

If none of the issues described above are of concern, then the interpretations by panelists are equally valued, and the discussion of the feedback should refine the understandings of the panelists about the use of the standard setting kernel. The discussion of the initial results will further help panelists come to a common understanding of the definitions resulting in translations that have similar meanings when panelists can modify their bookmark placements in the next round.

Another way to provide feedback is to show the locations of the estimated standards on the reporting score scale that correspond to the bookmark selections. Figure 7.2 provides an example of this type of feedback. Note that Figure 7.2 gives a different impression than Figure 7.1. The cluster of locations is at the right end of the scale, indicating that panelists indicated a fairly high standard by their bookmark placements. This is the case because the test was difficult, with the average examinee correctly answering less than half of the test items. The estimates in Figure 7.2 are more closely clustered than those in Figure 7.1 because there were several items with similar difficulties (and corresponding mapping locations) on the test so panelists could place bookmarks on different items but yield similar estimates for the standards.

As with the page number feedback, it is important to consider how providing information about the estimated standard for each panelist helps panelists produce an accurate translation using the standard setting process. Certainly, panelists can observe that their bookmark selections have resulted in different estimated standards. This is important feedback because it shows that the bookmark choices make a difference. However, the information in Figure 7.2, by itself, does not give information about how panelists might improve their translations if they believed they should. The results given in Figure 7.2 are on an abstract scale and panelists need to understand the connections between their selection of a bookmark location and the location of the estimated standard on the abstract scale. The important design issue

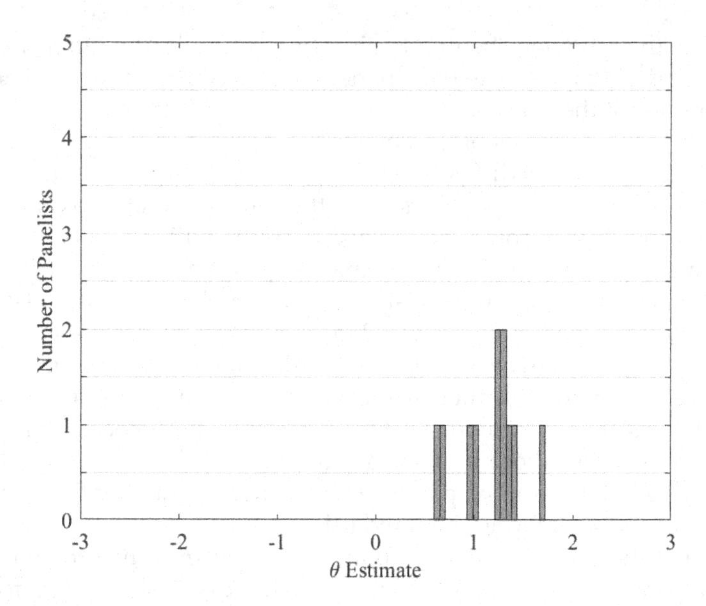

FIGURE 7.2
First round estimates of locations of standard

for a standard setting process is how to provide sufficient information to the panelists so that they understand the implications of the feedback and they know how to adjust the information they provide when making their translations using the standard setting kernel.

One approach could be to explain the item characteristic curves for each of the items and the mapping of the .67 probability to the IRT scale and then explaining how the panelists' bookmark placements are converted to points on the scale. In many cases, such an approach does not work well unless panelists have previous training in IRT. The usual schedule for a standard setting process does not have sufficient time available for a thorough short course on IRT and item mapping. A better approach may be to provide both the OIB page number feedback and the location of estimates feedback and explain the connection between the two.

Figure 7.3 shows a plot of the estimated standard against the midpoint of the bookmark placements. From that plot it is easy to determine how much change in estimated standard would occur if the midpoint of the bookmark placement were changed by some number of points. It is also clear from the graph that the panelist who placed the bookmark at page 40 in the OIB is an outlier from the group of panelists and that there are slight differences in estimated standards depending on the endpoints of the bookmark placements. Two panelists can have similar estimated standards while having different selections for their first and last bookmark placements.

The information in Figure 7.3 can be used to lead a discussion of the differences in bookmark placement relative to the information provided by the

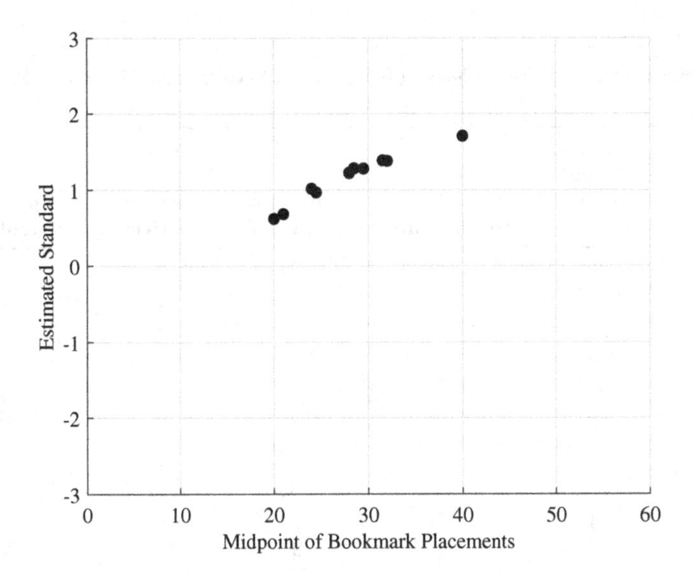

FIGURE 7.3
Relationship between bookmark placements and estimated standard.

definitions for the standard. Panelists can see their own locations on the graph and can see how much they would need to change bookmark locations to give an estimated standard at another location. The facilitator can help the panelists grasp the meaning of the information in the figure and explain how they can use it to decide on a modification to their bookmark placements if they believe that a change would result in a better translation of the definitions to a point on the scale.

Another type of information is often provided to panelists as feedback to help them determine where to place their bookmarks, that is, information about the observed difficulty of the test items in the OIB. This type of feedback is important because the panelists actively use judgments of the difficulty of items when making their bookmark placements. There are three different ways that the observed difficulty of test items is commonly reported. One way is the proportion or percent of correct responses for the full population of examinees. The second way is to estimate the proportion or percent correct for each item for persons who are located at the estimated standard based on the panelists' ratings. The third way is to show the locations where items have been mapped to the IRT scale when the OIB was developed. This is the location where the item has the probability of correct response used for mapping, .67 in the example, for a person with proficiency represented by that point on the scale.

Table 7.1 presents both the observed percent correct for the items in the OIB and the conditional percent correct for examinees at the median of the estimated standards, in this case 1.22 on the IRT θ-scale. The conditional percent correct has the influence of guessing on the item removed to be consistent

TABLE 7.1

Total Group and Conditional Percent Correct at a θ-Value of 1.22 for Test Items in the OIB

OIB Page Number	Total Group Percent Correct	Conditional Percent Correct	OIB Page Number	Total Group Percent Correct	Conditional Percent Correct
1	86	96	31	35	52
2	86	97	32	49	70
3	83	99	33	29	53
4	80	95	34	46	10
5	80	91	35	31	65
6	75	95	36	56	67
7	76	96	37	26	60
8	75	92	38	38	17
9	74	96	39	33	24
10	72	90	40	21	51
11	66	86	41	29	57
12	62	96	42	41	61
13	61	96	43	26	24
14	63	91	44	38	10
15	56	91	45	32	40
16	57	87	46	27	43
17	54	90	47	30	36
18	52	90	48	18	26
19	65	84	49	17	15
20	53	91	50	16	19
21	55	69	51	27	38
22	49	80	52	14	9
23	46	79	53	29	47
24	68	76	54	23	7
25	35	72	55	19	11
26	40	63	56	27	0
27	52	74	57	23	1
28	53	74	58	25	0
29	35	72	59	15	15
30	29	71	60	18	0

with the way the test items were ordered in the OIB. The total group percent correct is based on a sample of 5,000 simulated examinees and the values contain sampling variation. If another sample were drawn, slightly different values would be obtained. This is a partial explanation for the inconsistency of ordering of the total group percent correct and the OIB page number. For example, the sixth item in the booklet has a percent correct of 75, while the seventh item has a percent correct of 76. There are numerous examples of this kind of inconsistency in the orderings in the table. Besides sampling variation, the differences in ordering are due to differences in the data used for the ordering. The OIB ordering is based on the location on the θ-scale where the item has a probability of correct response after adjusting for guessing of .67. This location is based on the IRT model used to calibrate the test items. The total group percent correct is based on the full sample of data that is available and it does not make use of the IRT model.

The conditional percent correct is generally higher than the total group percent correct for the items because they refer to how persons at 1.22 on the IRT scale would perform on the test items instead of how the full group would perform that has a mean score on this scale of 0.0. The exception to that pattern of difficulty is for the difficult items at the later pages of the OIB. There the observed percent correct includes guessing while the conditional percent correct values are adjusted to remove the guessing effects. As a result, many of the conditional percent correct values are lower than those for the percent correct for the full sample.

As with all feedback, the important questions are what it is that panelists are supposed to take from the conditional percents correct, and how can they use the information to improve their translations. The answer to the first question is related to evidence that panelists are not particularly good at estimating the difficulty of test items and the percent of the items that minimally qualified examinees will answer correctly (Tinkelman, 1947; Clauser, Swanson & Harik, 2002; Clauser, Margolis & Clauser, 2014; Wyse, 2018). The expectation is that panelists will use information like that in Table 7.1 to refine their understanding of the difficulty of the test items. This should help them better estimate how the minimally qualified individual will perform on the test items.

As to the second question about how they can use the information, they should review their bookmark placements and see if their implied estimate of 67% correct for the minimally qualified examinee is reasonable for those items. If it is not, they should adjust their bookmark placements to provide better information for translating the definitions to the reporting score scale.

To give a better sense of how panelists might use the information in Table 7.1, some specific cases may be helpful. For example, one of the simulated panelists had the midpoint of the bookmark placements on page 20 of the OIB. The information in Table 7.1 shows that 53% of the full sample of examinees gave a correct response to the test item, and 91% of examinees at the group median standard responded correctly to the test item. The placement of the bookmarks by this panelist indicates that the interpretation of $\mathbf{minD}_C$ is that about 67% of the examinees who match that definition would respond correctly to items around page 20 in the OIB. There is nothing in the feedback given in Table 7.1 that is inconsistent with that panelist's interpretation. However, if there is discussion of the specific requirements of the test items and the capabilities described in $\mathbf{minD}_C$, then this panelist may understand how other panelists interpreted this information and gain a more refined understanding of the functioning of the test items and the nuances of $\mathbf{minD}_C$. This might result in the panelist rethinking the bookmark placements. Or it might result in the other panelists rethinking their bookmark placements. This implies that the feedback is important both as a check on the understanding of the relative difficulty of test items and a stimulus for discussion that can lead to greater understanding of the definitions and the functioning of the test items.

The third type of feedback that is commonly used with the Bookmark kernel is a mapping of the locations of the test items in the OIB on the IRT scale that is used for calibrating the test items. Table 7.2 provides one example of a table that shows the locations on the IRT scale where the items have a .67 probability of a correct response after adjusting for the estimated guessing parameters for the items. This table contains some important information for panelists about the clustering of item difficulties on the test and places where there are gaps in the distribution of difficulties of test items.

First, the table shows that there is a gap in the difficulty of items in the θ-range from .6 to .799. It is not possible for panelists to select a page from the OIB for their bookmarks that will result in a standard in that range. The table has cells shaded to show where the first bookmark placements were for the panel. The shading is darker when more than one panelist selected a page in the OIB. The shading indicates that four panelists placed their first bookmark before the gap and the rest placed the first bookmark after the gap. The majority of first bookmark placements were in the range 1.0–1.199.

The location of the second bookmark placements is indicated by the shaded block around the OIB page number. The locations of the second bookmarks range from page 24 to page 52. An important thing to note from the way the test items are distributed on the IRT scale is that the number of pages between bookmarks does not necessarily correspond to differences in location of the items. For example, suppose a panelist's first bookmark is on page 17 and the second is on page 28. The difference in the locations of those two test items is approximately .8 units in the IRT scale. The second bookmark is 11 pages later in the OIB. Suppose a second panelist places their first bookmark on page 28 and the second on page 40, 12 pages later in the OIB. The difference in locations on the IRT scale for these two items is .4 units. The second panelist has more pages between the two bookmarks than the first panelist, but the difference in location of the items on the IRT scale is only half as much. This is a result of the clustering of items at some of the locations on the IRT scale. This is essential information because if panelists want to adjust the bookmark locations to raise or lower their estimated standard, they need to move the bookmark beyond the cluster where it is located. Moving a bookmark from page 37 to 42 will make little difference in the estimated cut score because all those items are located near the same point on the IRT score scale. But a change from page 10 to page 15 will result in a notable change in the estimated cut score.

TABLE 7.2

Location of Test Items on IRT Scale for Bookmark Kernel

Lower Limit of Scale Range*	Page Number in the Ordered Item Booklet (OIB)					
–1.2	1	2				
–1.0						
–.8	3					
–.6	4	5	6			
–.4	7	8				
–.2	9	10				
0	11	12	13			
.2	14					
.4	15	16	17	18	19	20
.6						
.8	21	22	23			
1.0	24	25	26	27		
1.2	28	29	30	31		
1.4	32	33	34	35	36	
1.6	37	38	39	40	41	42
1.8	43	44	45			
2.0	46	47	48			
2.2	49	50				
2.4	51					
2.6	52	53	54			
2.8	55					
3.0	56					
3.2	57	58				
3.4						
3.6	59					
3.8						
4.0						
4.2						
4.4						
4.6	60					

* Note: The range specified by the first column is from n to $n + .199$. For example, from 1.0 to 1.199 followed by 1.2 to 1.399.

As with other types of feedback, an important consideration is how panelists can use the information that is provided to improve the translation of the definitions into a point on the reporting score scale. The review of the distribution of item locations facilitates understanding of the probabilistic nature of the item response data. When there are clusters of items at points on the IRT scale, thinking about the point where the probability of a correct response first drops below .67 is conceptually the same as the point where less than .67 (or 2 out of 3) of the items at that point will have a correct response. This may be easier for panelists to comprehend than the abstract probability value of .67. It may also be easier to compare sets of items to the definitions than single items. Of course, it is important that the facilitator clearly explain the information about test item locations along with options for how panelists can use the information.

7.2.2 Feedback about the Consequences of the Use of the Estimated Cut Score

In some designs for a standard setting process, feedback is given on the results of applying the estimated cut score to the target population of examinees. This type of feedback is given different labels, including "consequences" and "impact" feedback. This feedback is given so that panelists can determine whether the translation of policy to a point on the reporting score scale results in a proportion above the cut score that matches the intent of the $\mathbf{D_P}$. If the $\mathbf{D_P}$ uses language like "high standards" or "rigorous standards," the expectation is that a relatively small proportion of individuals will exceed the standard. If the language of the $\mathbf{D_P}$ uses terms like "minimum competency" or "basic skills," the expectation is that there will be a high proportion above the translation of $\mathbf{D_P}$ to the reporting score scale. Of course, this type of feedback can only be provided if the performance distribution for the target population is available at the time of the standard setting meeting.

In some cases, $\mathbf{D_P}$ specifies multiple standards. For example, NAEP has three defined levels: Basic, Proficient, and Advanced. Even those simple labels for the levels of performance have meaning related to the proportion of examinees that are expected to be above the corresponding points on the reporting score scale. When there are multiple levels specified by $\mathbf{D_P}$, the proportions within each level or above each point on the scale may be provided as feedback.

There is a concern when providing consequences feedback that panelists will give it more consideration than the information in the definitions. That could result in a mismatch between the point on the reporting score scale panelists recommended and the information given in $\mathbf{D_C}$ and $\mathbf{minD_C}$. For that reason, some standard setting processes do not provide consequences feedback at all, or this type of feedback is given toward the end of the process so it will not have excessive influence on the results.

In all cases of providing consequences feedback, it is important to consider how the feedback will help improve the translation of $\mathbf{D_P}$ to a point on the reporting score scale, and how panelists are expected to use the information. The most common design for a standard setting process is to provide feedback, encourage discussion among the panelists, and then give them an opportunity to revise the information they provided using the standard setting kernel. For those activities to result in an improvement in the translation of $\mathbf{D_P}$ to the point on the reporting score scale, panelists must know how to use the feedback to adjust the information they provide from the use of the standard setting kernel.

The example from Section 7.2 provides a useful context for showing how panelists can use consequences feedback to improve the translation. That example was for a placement test into a mathematics course. The language of the definitions was not presented in Section 7.2, but an example of a $\mathbf{D}_C$ is "should have a HIGHLY DEVELOPED understanding of algebra, a SOLID knowledge of trigonometry, and the related manipulative skills" [capital letters in the original]. This definition was taken from the Michigan State University website for placement into the calculus course for those who will make "serious use of calculus" in their majors (https://math.msu.edu/prepare_for_math_at_msu/default.aspx#Placement). Note that this definition includes language that suggests that this should be a high standard.

If the example given in Section 7.2 is for a placement examination that is consistent with this type of $\mathbf{D}_C$, the simplest type of consequences feedback that could be provided to panelists is the percentage of examinees that exceed the estimated point on the reporting score scale. For the example, using the median of the estimated points on the score scale for the panelists, the percent is 10.66. If given this information, the panelists then have the task of determining if this percentage for students who take the placement test is reasonable given their understanding of $\mathbf{D}_C$. If a panelist decides that that percent above the estimated standard is not consistent with $\mathbf{D}_P$, $\mathbf{D}_C$ or $\mathbf{minD}_C$, then the panelist needs to decide how to change the information they provide using the tasks specified by the standard setting kernel to yield a result that is consistent with the definitions. Often, however, it is not at all clear how the percent passing will change with a change in bookmark placement. Will moving the bookmark one page in the OIB result in a big change or a small change in the percent above the estimated point on the reporting score scale?

Figure 7.1 gives some information about the result of moving the bookmark. However, that figure relates that bookmark placement to the estimated standard, not the percent above the standard. An alternative is to provide the panelists with the percent passing for each possible bookmark placement in the range of OIB pages that they are considering. For this example, that range is from OIB pages 10–52 (see Table 7.2). Table 7.3 provides the percentage above the point on the reporting score scale implied by each OIB page for that range. The highlighted cells show the median bookmark page representing the estimated standard for the full group of panelists.

The task for the panelists when receiving this information is to determine if the percent who would receive advanced placement matches the implications of the definitions. For the sample of students taking the advanced placement test, do only 11% have a highly developed understanding of algebra and solid knowledge of trigonometry? If a panelist believes that the proportion above the estimated standard should be different, what should he or she do to get a better translation of policy? If a panelist's understanding of $\mathbf{D}_C$

TABLE 7.3

Percent Above Standard Estimated from Bookmark Placements on Pages in the OIB

OIB Page Number	% Above Standard	OIB Page Number	% Above Standard	OIB Page Number	% Above Standard	OIB Page Number	% Above Standard
10	51	21	18	32	7	43	4
11	49	22	17	33	7	44	3
12	44	23	16	34	7	45	3
13	43	24	15	35	7	46	2
14	37	25	12	36	7	47	2
15	31	26	12	37	5	48	1
16	30	27	12	38	5	49	1
17	30	28	11	39	5	50	1
18	29	29	11	40	4	51	1
19	29	30	10	41	4	52	1
20	29	31	9	42	4		

Note: Highlighted information is for the median midpoint of the bookmark placements.

suggests that the percent above the estimated standard should be higher, then he or she should move the bookmark earlier in the OIB. However, Table 7.3 shows that moving the bookmark one to three pages earlier, will change the result by only 1%. That is because there are several test items of similar difficulty to the one on page 28 and the bookmark would have to be placed on page 24 for the percent above the standard to reach 15%. Similarly, if the panelist believes the percentage above the standard should be smaller, then moving the bookmark one page will have no effect at all. The table shows how much the bookmark would have to be moved to reach the desired percentage. This feedback is useful for showing panelists how much change in bookmark placement is needed to achieve their desired results.

There is a complication to providing this information that shows possible inconsistencies in panelists and policy makers' understandings of the language of the definitions and the functioning of the test items. It may be that panelists' understandings of $\mathbf{D}_P$ lead them to change their bookmark placements to result in a different percent above the standard, but then the items that are near the bookmark placement are no longer consistent with $\mathbf{D}_C$ and $\mathbf{minD}_C$. In a perfect world, such inconsistencies would not occur. But, in actual standard setting studies, the definitions are less precise than would be desired. Part of the reason for this is a need to communicate in a concise way to a variety of audiences. Another reason is that members of the agency may not be experts in the subject matter of the construct implied by the $\mathbf{D}_P$. The coherence of the definitions depends on the fidelity of translations of $\mathbf{D}_P$ to $\mathbf{D}_C$ and the test design that operationally defines the construct that defines the continuum for the location of the estimated standard. The standard setting process has many component parts, and it is difficult to have a high level of consistency among all of them.

The critical factor when giving consequences feedback is getting panelists to actively consider the differences between the implications of the definitions for both the proportions exceeding a standard and the descriptions of capabilities of those who exceed the standard. If those components of the process are inconsistent, there needs to be a means for resolving the inconsistency.

7.2.3 Other Types of Feedback

Whole-Booklet Feedback. The previous two sections of this chapter give examples of feedback for a standard setting process using a Bookmark kernel. Each standard setting kernel has feedback that is particularly useful for helping panelists refine the information they provide for translating the definitions into a point on the reporting score scale for a test. For example, if the test operationalizing the construct implied by the $\mathbf{D}_P$ is composed of a set of essay prompts, the kernel used for the standard setting process might be having panelists specify the mean score on each essay for each prompt for persons who are described by $\mathbf{minD}_C$. This kernel is sometimes labeled Extended Angoff or Mean Estimation (Hambleton, 1998). One approach to estimating the standard is to sum the mean values specified by a panelist for each of the essay prompts. A type of feedback that could be used in this case is the actual set of essays produced by candidates with scores located at the proposed standard. This is sometimes labeled as whole-booklet feedback. Panelists can review the work of those candidates and determine if it matches the information given in the definitions.

As with other types of feedback, information needs to be provided with whole-booklet feedback to help panelists understand the implications of the feedback for the translation to the point on the reporting score scale. Examinees are seldom consistent with the way they respond to multiple essay prompts, so the minimally qualified examinee will likely have variation in the quality of their work in the booklet. Also, the booklets are selected based on the scores given to the essays by trained readers. Panelists will need to understand the scoring process, so they can understand how readers assigned the scores and the amount of error in the scoring.

Ultimately, the goal of the whole-booklet feedback is to help panelists adjust their estimates of mean performance on each essay to better translate the information in the definitions. If they decide that the performance on the essays in a booklet are not of sufficient quality to match the descriptions in the definitions, how much should they change their estimates to get a set of booklets that do match the definitions? To help the panelists, it may be useful to give examples of booklets that are from several points above the estimated standard and from several points below. Then panelists will be able to note the amount of change in quality of the essays that match the differences in scores on the reporting score scale.

TABLE 7.4

Item-Score Strings for Two Examinees at the Estimated Bookmark Cut Score

Page Numbers for Test Items in the OIB					
1-10	**11-20**	**21-30**	**31-40**	**41-50**	**51-60**
1111111110	1111111110	1111111100	1111110101	0100011101	0001010100
1111111111	1111111111	1111111001	1101101111	1000111011	1000000010

Lewis and Haug (2005) used another version of whole-booklet feedback for standard setting with the Bookmark kernel. They provided panelists with examples of the actual strings of 0s and 1s for the scores on the test items for examinees that had performance estimates that were close to the estimated cut score. They provided this feedback to show that examinees do not answer all the test items before the bookmark correctly and answer all the test items after the bookmark incorrectly. Table 7.4 shows two item-score strings for the bookmark example given in the previous section. The median of the bookmark placements was on page 28 in the OIB. The examples show that most of the test items before the item on OIB page 28 were answered correctly, but there were some exceptions. Also, there were a number of correct responses on test items in the OIB after page 28. This test was composed of multiple-choice test items, and it is likely that some of the correct responses were due to guessing. If panelists expected that the number correct score corresponding to the bookmark placement would be 28, these results show that was not the case. The two examinees had number correct scores of 42 and 44, respectively.

The main question related to this type of whole-booklet feedback is about how panelists are expected to use it to improve the translation of the definitions to the reporting score scale.

Reckase Chart. Common types of feedback to panelists when the Angoff probability estimation kernel is used are the proportion correct for the items on the test for the target population of examinees and the conditional proportion correct for examinees near the estimated cut score. This is essentially the same information provided in Table 7.1, except items are listed in the order administered to the examinees instead of the order in the OIB. This information is provided so that panelists can determine if their estimates of probability of correct response for examinees matching $\mathbf{minD}_C$ are reasonable and consistent with the way examinees actually perform on the items. As with other types of feedback, panelists need to determine how to use this feedback to improve the information they provide using the standard setting kernel. For the Angoff probability estimation kernel, after receiving the feedback, panelists can modify their initial probability estimates for items to improve the translation to the reporting score scale. However, understanding the likely result of changes to the estimates is challenging, especially if the reporting score scale is based on IRT. To help panelists understand how to change their probability estimates, a feedback mechanism was developed for the standard setting process for NAEP that was labeled by Loomis, Hanick, Bay, and Crouse (2000) as a Reckase chart. This chart shows the conditional proportion of correct responses at possible cut scores on the test. To help panelists understand the information in the chart, panelists are told to indicate their probability estimates on the chart as well as their estimated standard.

Table 7.5 gives an excerpt from a Reckase chart for the same items that were used for the Bookmark kernel example earlier in this chapter. The full chart would contain the full range of possible reported scores and all the test items. The full range of the reported scores is roughly from –3 to 3: the usual range of IRT-based estimates of proficiency. This excerpt contains the first ten items on the test and the range of proficiency estimates (0–2) that contains the estimated standard. The item number in the table refers to the number in the test booklet that was administered to examinees. In that booklet, items

TABLE 7.5

Reckase Chart for First Ten Items on Test for the Region around the Estimated Standard

Reporting Sore Scale	Item Number									
	1	2	3	4	5	6	7	8	9	10
0	.79	.34	.57	.12	.16	.53	.84	.54	.42	.18
.1	.81	.34	.58	.12	.16	.55	.86	.58	.42	.18
.2	.84	.34	.60	.12	.16	.57	.87	.62	.42	.18
.3	.86	.34	.61	.12	.16	.59	.89	.66	.42	.18
.4	.88	.34	.62	.13	.17	.61	.90	.70	.42	.18
.5	.89	.34	.63	.13	.17	.63	.91	.73	.42	.18
.6	.91	.35	.64	.13	.18	.65	.92	.77	.42	.18
.7	.92	.35	.65	.14	.18	.67	.93	.80	.42	.18
.8	.94	.36	.66	.14	.19	.68	.94	.82	.42	.18
.9	.95	.37	.67	.15	.20	.70	.94	.85	.42	.18
1.0	.96	.39	.68	.16	.21	.72	.95	.87	.43	.18
1.1	.96	.41	.69	.17	.22	.73	.95	.89	.44	.18
1.2	.97	.44	.70	.19	.23	.75	.96	.90	.46	.18
1.3	.97	.47	.72	.20	.25	.76	.96	.92	.51	.18
1.4	.98	.52	.72	.22	.27	.78	.97	.93	.60	.18
1.5	.98	.57	.73	.25	.29	.79	.97	.94	.72	.18
1.6	.99	.63	.74	.28	.31	.80	.97	.95	.83	.18
1.7	.99	.69	.75	.31	.33	.82	.98	.96	.91	.18
1.8	.99	,74	.76	.35	.36	.83	.98	.96	.96	.18
1.9	.99	.80	.77	.39	.39	.84	.98	.97	.98	.18
2.0	.99	.85	.78	.43	.42	.85	.98	.98	.99	.18*

Note: A panelist's Angoff ratings are highlighted with the gray background. *The actual rating by the panelist was .26.

were not ordered in difficulty, so this is not like the OIB that is used with the Bookmark kernel. The highlighted cells in this table show a panelist's probability ratings for each item.

There are several features about the test items and the panelist's ratings that can be noted from this table. First, Item 1 is an easy item. Even those in the middle range of the reporting score scale are expected to answer it correctly about .80 of the time. The panelist is aware that it is easy and rates it .99 expecting that about everyone will answer it correctly. The chart is showing that .99 is the upper asymptote of the test item and high-performing examinees do indeed answer it with that level of accuracy.

In contrast to Item 1 is Item 10. This is a very difficult test item that only the highest performing individuals answered correctly. The panelist noted the difficulty and gave a rating that was slightly above the random guessing value for the item, but that was still higher than the observed rate of correct responses for most examinees. The ratings for Item 1 and Item 10 are outliers. These were extreme items, and the panelist did not accurately judge how extreme they were. One of the functions of the information in the table is to inform the panelist about the functioning of the items so they can make more informed judgments.

The ratings for the other eight test items in the table are clustered around the estimated standard that is about 1.2. The variation around the 1.2 row of the table shows the variation in judgments by a panelist. This is a typical pattern of Angoff kernel ratings. Panelists cannot be expected to be consistent with the IRT calibration. This type of feedback can give additional information that can suggest test items that panelists should give more thorough study. After seeing this feedback, panelists will usually reduce the variation in ratings around the row, but most do not pick a row of the table and change all their ratings to match the numbers in the row.

7.3 Auxiliary Information

Some standard setting processes provide information to panelists that neither concerns their understanding of the standard setting kernel nor the direct consequences of the cut score that is estimated from the information panelists provide. This could be information from other testing programs, the cut scores set for other grades, and how groups other than the target group would perform relative to the cut score that has been set. The goal of providing this information is to give panelists a broad perspective on what is realistic for the standard. There is a wide variety of feedback of this type and new options are appearing as those who design the standard setting processes become more creative or try to address problems that arise during a specific standard setting process. Several examples will be presented, but there is no intention to be comprehensive. Rather the goal is to identify basic principles.

Results from Other Tests. It is relatively common for the design of the tests that operationalize the construct implied by the policy definition to be revised. Over time, the skills and knowledge that are the focus of instruction change or there are new requirements for jobs. Of course, when this happens, there should be new $\mathbf{D_C}$s for the standards and a new standard setting process conducted to estimate the cut score on the reporting score scale for the revised test. Under these circumstances, the agency that calls for the standard may want to make sure that the standards are consistent over time. That is, the $\mathbf{D_P}$ may not change and there is an expectation that the proportion exceeding the estimated cut score will not change very much.

One approach to addressing this policy concern is to have panelists begin the standard setting process with information about the proportions exceeding the standard on the previous version of the test and the changes to the test design. In a sense, panelists are being asked to edit a previous translation of policy to the reporting score scale rather than create a new translation. This may work reasonably well if the test design has not changed in any dramatic way, the $\mathbf{D_C}$ are produced to be consistent with the previous $\mathbf{D_P}$, and the new reporting score scale has similar meaning to the previous score scale. However, the panelists may consider this information as an attempt to tell them what the translation should be rather than asking them to develop a good translation on their own.

From the perspective of the theory of standard setting presented here, providing the information from the previous test is equivalent to modifying $\mathbf{D_P}$ to include the need for consistency with the previous translation. It also implies that the $\mathbf{D_C}$ and $\mathbf{minD_C}$ were developed to be consistent with the modified $\mathbf{D_P}$. Further, the test design for the new/revised test that operationalizes the construct implied by the modified $\mathbf{D_P}$ must be consistent with all the definitions. The analogy to language translation is that a document has been translated to another language and that original document has been modified in some way. If the details of the modifications are clear, using the original translation as a starting point will make developing the translation of the modified document more efficient. But the efficiencies will only be present if the modifications are clearly specified, and they are not too extensive. If the modifications are extensive, it would be better to do a totally new translation. The same is true for standard setting on tests. Providing the earlier results is helpful if the modifications to $\mathbf{D_C}$ are not too great.

Results from Other Grades. In the educational testing environment, it is common to administer achievement tests in subject matter areas to students at several grade levels with the goal of documenting improvement in skills and knowledge over grades. In such cases, the reporting score scales for each of the grade level tests are linked to form what is called a "vertical scale," meaning a scale that spans a wide range of performance. Such a scale can be used for reporting the performance of students from multiple grades. This type of scale allows users to note the differences in performance of students as they progress through the educational system.

When standards are set on the achievement tests for each grade, for example to define how much achievement is needed to be labeled as proficient, the pattern of increase for the points on the vertical scale defining proficient can be noted. In some cases, there may be what are considered anomalies or reversals in the ordering of the standards over grades. For example, the proficient point on the vertical scale for fourth-grade students is higher than the proficient point for fifth-grade students. To address this issue, standard setting processes that use results reported on vertical scales sometimes use a process that is called "vertical articulation" (Cizek & Bunch, 2007). Cizek and Bunch (2007) also use the term "vertically moderated standard setting (VMSS)" to indicate that the standard setting process provides information about the relationship between standards estimated at different grades.

When the standard setting process includes vertical articulation or vertically moderated standard setting, panelists first perform the tasks specified by the standard setting kernel to provide information needed to get an estimate of the point on the vertical scale that corresponds to $\mathbf{D}_P$ and $\mathbf{D}_C$ for their grade. Separate panels are convened for each grade so the panelists will be familiar with the student characteristics at each grade. Then the panelists for each panel are provided with information about where the standards were set at other grades and are asked to consider that feedback before making a final recommendation for the location of the standard for the grade that is their focus.

As with all standard setting procedures, there are many different ways to provide and use this information. Sometimes, all panels receive the information. Other times, individuals are selected from panels at adjacent grades to form a new panel for the specific purpose of vertical articulation. The reason for the joint panel is to get expertise about the content and instruction for adjacent grades. Sometimes all panels are combined to form a large group to review the full set of recommended standards across all the grades.

Lewis (2002) and Lewis and Haug (2005) indicate that including vertical articulation in the standard setting process implies that there is information in the $\mathbf{D}_P$ requiring standards to be monotonically increasing over grades. They suggest several relationships between grade level and the location of the standard on the vertical scale. They include linear and nonlinear relationships as options for the increase in location of standards with the increase in grade. In some cases, they make projections of what locations on the vertical scale would be if the increase from grade to grade took a particular mathematical form. After receiving that information as feedback, panelists can adjust the locations of the standards before making a final recommendation.

7.4 Timing of Feedback

One of the decisions that designers of standard setting processes need to make is when to give feedback or ancillary information. Generally, process feedback is given after every round of the standard setting process so that panelists will improve in their understanding of the tasks required by the standard setting kernel. Normative feedback tends to be given later in the process because of a belief that it will have less impact after provisional standards have been set based on content definitions. The reason for wanting to limit the impact is that $\mathbf{D_P}$ usually specifies a criterion-referenced standard rather than a norm-referenced standard. However, there is little research that shows the consequences of giving normative feedback at different points in the standard setting process. As early as Shepard (1980), there was a call for more research on the timing and impact of feedback. It would be useful to have a carefully controlled study with normative feedback given after the first, second, and third rounds of the standard setting process with randomly equivalent groups of panelists for each condition to determine the influence of it on the timing on the estimated cut scores.

The little research that exists (see Reckase (2000) for a summary of the work done for NAEP standard setting) indicates that the effect of feedback diminishes when it is presented after multiple rounds of application of the standard setting kernel. Panelists seem to identify a location for the estimated cut score, and their confidence in it gets stronger with each round of rating. They are reluctant to refine their judgments once they have a reasonable level of confidence in the translation to the reporting score scale that they have produced.

7.5 Impact of Feedback on the Standard Setting Process

There are several studies that have looked at the impact that feedback has on the results of a standard setting process. Generally, as suggested by the theory of standard setting presented here, feedback helps panelists refine the information that they provide for translating the definitions to a corresponding point on the reporting score scale. Results reported by a number of researchers (Clauser, Mee & Margolis (2013); Cizek & Bunch (2007); Hambleton (1998); Shepard, Glaser, Linn & Bohrnstedt (1993); Yang (2000)) show that the estimated standards have reduced variation after feedback was provided. This is consistent with the theoretical position that there is a "true standard" like a "true score" that is unobservable but that is estimated from the information provided by the panelists. As panelists improve their understandings of the tasks required by the standard setting kernel and the nuances of $\mathbf{D_P}$, $\mathbf{D_C}$, and $\mathbf{minD_C}$, they should all be providing information for estimating a similar point on the reporting score scale.

Clauser, Mee, and Margolis (2013) add a caution to the generally positive research on the use of feedback in standard setting studies. They emphasize that it is important that panelists have a firm grasp of the meaning of the feedback. If panelists misinterpret the meaning of the feedback, for example interpreting the mapping probability for the Bookmark kernel of .67 not as the conditional probability of correct response but as the proportion correct on the test, panelists may be consistent in the bookmark placements they provide but inaccurate in their translation of the standard. Clauser, Mee, and Margolis (2013) recommend checking for bias in the estimates as well as agreement or consistency. While Hambleton (1998) noted that the changes to estimated standards after feedback tend to be small, it is an important question whether the changes improve the connection between the definitions and the corresponding point on the reporting score scale or cause bias in the estimates. This is one of the concerns that leads designers of standard setting processes to give consequences or impact feedback after the second or third round of a standard setting process. Designers of the process worry that panelists will have in mind a percentage of examinees that should exceed the standard and change the information required by the standard setting kernel to yield that preconceived percentage. In doing so, they may be distorting the connection between the definitions and the point on the reporting score scale that is supposed to correspond to them. This is a major concern when $\mathbf{D_C}$ and $\mathbf{minD_C}$ are reported along with the cut score.

7.6 Conclusions and Recommendations

Any translation task is cognitively challenging. When learning to translate from one language to another, a student usually has an instructor who gives assignments and then reviews/corrects them and gives feedback so that the student knows when he or she has performed the task well. The task of translation from $\mathbf{D_P}$ and $\mathbf{D_C}$ to a point on a reporting score scale is at least as challenging as language translation. Persons with many relevant skills are selected to participate in the standard setting process, but they need feedback about their performance on the required tasks to help them learn the nuances of the task. It is interesting to imagine a short course on the tasks for a particular standard setting kernel with instruction on the details of the tasks, examples, and assignments designed to give persons practice and feedback about their skills in doing the task. Only those who pass the short course would be selected to do the standard setting.

Of course, such a short course is impractical for many reasons. Instead, a standard setting process is scheduled for a day to a week, possibly with materials sent to participants in advance to give them the context for the work they are to do. The expectation is that the panelists will understand the tasks specified by the standard setting kernel well enough after some brief explanations and possibly some practice tasks for them to give useful information that can be aggregated over a number of panelists to get a good estimate of the standard. The fact that several rounds of the application of the standard setting kernel are typically used in a standard setting process is an acknowledgment that the brief training on the standard setting kernel is not sufficient to get a quality translation. In analogy to the translation of text from one language to another, the result after the first round of a standard setting process is a rough draft. Then panelists need feedback to help them evaluate the quality of their own work and give them guidance about how they can improve the translation to the reporting score scale.

This chapter describes many different kinds of feedback. Some of the feedback is about the judgments required by the standard setting kernel and panelists' levels of understanding of the judgment process. Other types of feedback give other relevant information about the performance of the examinee population. An important part of the design of the standard setting process is determining the kinds of feedback to include. Some of that design process deals with policy decisions about whether normative information and performance on other tests should be included. These important issues should be resolved early in the design process.

Finally, feedback is helpful only if panelists understand its meaning and relevance to the next steps in the process. I have observed standard setting processes where several types of feedback are presented to panelists after a round of ratings and there is not sufficient time to explain the meaning of the feedback and how it can guide judgments in later rounds. In such cases, panelists tend to ignore the feedback. For feedback to be effective, panelists must gain a thorough understanding of the meaning of the feedback and have some information about how the feedback can guide their next judgments. It is important that reports of standard setting processes indicate how the design of the process will help panelists understand all aspects of the process, including what to do with the information provided by feedback. Mitzel et al. (2001) indicate that when panelists do not understand the feedback or are overloaded with it, they will revert to using heuristics that might induce bias in the estimates of the standards. They emphasize training about feedback as a way of helping panelists understand the feedback and how to use it.

8

Estimating the Standard

8.1 Introduction

It is surprising that the idea of estimating something from data can be controversial. But in the case of deciding on the point on a reporting score scale for a test to use when making decision, even the use of the term "estimation" is controversial. This position was presented by Shepard, Glaser, Linn, and Bohrnstedt (1993) in the context of performance standards for the National Assessment of Educational Progress (NAEP). Their position was as follows:

> Because standard-setting methods were developed in the psychometric literature and involve numerical calculations, they are often imbued with a false sense of scientific precision. In reality, measurement experts acknowledge that no true standard exists in nature for standard-setting techniques to "estimate." Therefore, a standard-setting method, even when implemented precisely as recommended in the literature, will not necessarily produce a true or valid standard. It will only summarize the judges' opinions.
>
> **(p. 23)**

In other parts of the same report, they recommend against reporting standard errors for the results of a standard setting process because such reports make the procedure seem more rigorous than they believe that it is.

There is a certain irony in the position taken by Shepard, Glaser, Linn, and Bohrnstedt (1993) because in another part of the same document they state that the standards that were set on the NAEP reporting score scale were "set unreasonably high." To say that a standard is too high, there must be a sense of an appropriate level for the standard and the observed value exceeds that level. How do they determine the appropriate level for comparison? In this case, the appropriate level was derived from other studies that obtained estimates of proportions of students above standards. Somehow it was determined that these other estimates were better than those obtained through a formal standard setting using the Angoff probability-estimation kernel.

DOI: 10.1201/9780429156410-8

Clauser, Clauser, and Hambleton (2014) take a similar position, but one that is more nuanced concerning estimation.

> It is important to note that viewing the Angoff method as an estimation procedure does not imply that an absolute standard somehow exists in nature waiting to be discovered. Instead during the training process each judge relies on his or her expert judgment to internalize the ability level associated with the minimally capable examinee. This value can be unique for each judge but should be constant for that judge within round of judgements. The individual judgments provide evidence for estimating the judge's expert opinion so that it can be translated onto the score scale.
>
> **(p. 22)**

In another article, Clauser, Margolis, and Clauser (2014) give a more technical interpretation of this position using a generalizability theory framework. That framework assumes that there is a universe of items and panels, and judges nested within panels.

> To estimate the cut score item-level judgments are averaged (or summed) across items within judges and then averaged across judges (and possibly across panels). This grand mean (across items, judges, and panels) typically is used as the estimated cut score (although other approaches are possible). The purpose of the generalizability theory analysis is to estimate the error associated with this grand mean.
>
> **(p. 131-132)**

These descriptions of the estimation method suggest that there is a parameter that is being estimated and it is the grand mean of the standards based on the opinions of an appropriate population of panelists. This would be the equivalent of the grand mean of heights of a defined population of individuals. There is no need for the heights to be the same. It is expected that there will be a distribution of heights, but there is a parameter to be estimated that is the population mean of the distribution. So it is when thinking about estimation of a standard in the context of generalizability theory. Panelists will have different opinions about the location of the standard, but the parameter to be estimated is the grand mean of those opinions. The observed sample will be different than the population value because of sampling error and the magnitude of contributing factors to the sampling error can be estimated through generalizability analyses.

The position taken in this book is that the data that are collected from panelists when they perform the activities specified by a standard setting kernel are used to *estimate* a point on the reporting score scale that corresponds to the intended standard implied by the policy definition $\mathbf{D}_P$ produced by the agency that calls for the standard. Analogous to the true score in classical test theory, there is a true standard that is unobservable, but that can be estimated from information collected from panelists. Just as the standard error of measurement can be computed to give a sense of how close an observed score is to the true score, the amount of error in the estimate of a standard can be computed to give information about the level of confidence that can be placed in the estimate of the location of a standard on the reporting score scale.

This conception of estimation and standard setting is not as far from that of Shepard, Glaser, Linn, and Bohrnstedt (1993) and Clauser and colleagues as it may seem. The standard is still a matter of opinion as stated by those authors. But it is the opinion of those in the agency as expressed in the $\mathbf{D}_P$. The true standard is the point on the reporting score scale that is most consistent with $\mathbf{D}_P$. The standard is not an estimate of the opinions of the panelists who have the role of translators of $\mathbf{D}_P$ to the point on the reporting score scale, especially if the panelists have a different opinion than the agency. This conception of estimation as part of standard setting is consistent with the usual practice of considering the estimated standard as a recommendation to the agency and that the agency can adjust the standard if the members believe that the recommended standard does not match their policy. Of course, the agency could decide that their policy is that the standard is an "average" of the opinions of a qualified group of individuals such as the members of their organization. In that case, the agency should not adjust the standards unless there was evidence that there were technical problems in computing the estimated standard such as poor sampling or computational error.

The general position that guides the work presented here is that the agency has produced a policy definition $\mathbf{D}_P$, which describes a level of competency on a construct of interest. The $\mathbf{D}_P$ can be elaborated to describe the specific knowledge and skills needed for the desired level of competency as presented in $\mathbf{D}_C$ and there is a test that has been designed to measure the skills and knowledge given in $\mathbf{D}_C$. This test is an operational definition of the construct implied by $\mathbf{D}_P$. Further there is a minimally acceptable level of performance on the skills and knowledge in $\mathbf{D}_C$, $\mathbf{minD}_C$, and the point on the reporting score scale for the test that corresponds to that level of performance is the desired standard. Under the ideal conditions that $\mathbf{D}_P$, $\mathbf{D}_C$, and $\mathbf{minD}_C$ perfectly communicate intentions to all the parties involved, and the test provides reported scores that are a perfect representation of the intended construct, then the result is the point on the reporting scale that is the true standard, ζ_{Ag}, desired by the agency. It is this value that is the goal of estimation in the standard setting process.

Of course, this ideal situation never exists. The definitions do not provide perfect communications of desired competencies. The test is not a perfect representation of the desired construct, and it does not have perfect reliability of measurement. And those asked to participate in the translation process from $\mathbf{D}_P$ to ζ_{Ag} cannot do this without error. All these components are sources of error in the process, and it is the goal of a standard setting process to perform this translation in a way that minimizes the error.

Before considering the details of the estimation of the standard for various methods, the assumptions behind the estimation will be summarized. A common set of assumptions apply to most of the methods. Listing them emphasizes the common threads among many of the methods.

8.2 Assumptions Underlying the Estimation Processes

The descriptions of standard setting procedures seldom mention the assumptions that are made when computing the cut score for a test. For example, when describing the extended Angoff kernel, Plake (1998) indicates that panelists give the anticipated score on open-ended items scored from 1 to 4 and a weight that represents the relative importance of each to the "overall purpose of the assessment." The estimated standard is a weighted aggregation of the scores on the individual tasks.

This example was not selected because it was an unusual case. Most reports of the estimation of the standard using a standard setting kernel give similar descriptions. But these descriptions do not give any information about the assumptions used when computing the estimate. In this example, there are two basic assumptions that are being made. The first is that there is a meaningful continuum that is defined by the reporting score scale for the test. The second is that the information that is combined to form the estimate consists of independent evaluations of the performance of the minimally qualified examinee on the tasks in the test. The next sections of this chapter describe these two assumptions, and variations to them.

8.2.1 Reporting Score Scale

The theory presented in this book is based on the idea that an agency calls for a standard and defines the context of the standard with a policy definition $\mathbf{D_P}$ that at least implies a continuum for the concept that is the target for the standard. The continuum is abstractly represented by ζ and the point on the continuum that corresponds to the intended standard is represented by ζ_{Ag} with the subscript Ag indicating that it is the standard intended by the agency. Persons to be assessed relative to the standard are also assumed to have locations on the continuum and those locations are represented by ζ_j. The decision that is made relative to the standard is whether $\zeta_j > \zeta_{Ag}$. This implies that the scale represents the magnitude of a construct implied by $\mathbf{D_P}$ and the goal is to determine if individuals exceed the magnitude specified by the standard, ζ_{Ag}.

The minimal assumption needed for estimating ζ_{Ag} is that there is a scale that orders candidates on the construct specified by $\mathbf{D_P}$ and $\mathbf{D_C}$. In many cases, this scale is assumed to meet a unidimensionality assumption as specified by the concept of essential unidimensionality defined by Stout (1987).

This concept allows minor dimensions if there is a dominant dimension that increases its dominance as the test length increases. In most cases, unidimensional item response theory models can be used to construct the reporting score scale if essential unidimensionality holds.

The assumption that there is a scale that orders candidates does not require essential unidimensionality. Standards can also be estimated on scales that are formed as weighted composites of multiple component parts. This is an implied assumption for many tests of proficiency that have strong content representation requirements. The reporting score scale is the sum of component parts. These reporting score scales are considered as linear compensatory scales because high performance on one component part can compensate for low performance on another component part. Often, the weighting is implemented through the item selection process. Specified numbers of items are included in the test for each defined area of the content domain. Computerized adaptive tests (CATs) implement the same implied weighting through what is called "content balancing"—items are selected so that specified proportions of items come from each content domain.

In rare cases, the weighting of component parts of a test is made explicit. Plake (1998) describes a procedure with a standard setting kernel called Judgmental Policy Capturing (Jaeger, 1995) where panelists evaluate profiles of scores on parts of an examination system using a 1–4 scale. The evaluations of the panelists are then predicted from the scores on the parts using linear regression. The point on the predicted score scale that best divides the acceptable profiles from those that are not acceptable is the estimated standard. The weighing of the parts comes directly from the regression analysis. In a sense, the statistical analysis is "capturing" the panelists intended weighting of the parts of the assessment system.

8.2.2 Independence of Information from Panelists

Another basic aspect of the theory presented in this book is that panelists are translators from the language of policy to the language of the reporting score scale for the test that is used to operationally define the construct that is implied by the policy. However, because few persons are proficient in both the language of the definitions ($\mathbf{D}_P$, $\mathbf{D}_C$, and $\mathbf{minD}_C$) and the language of the reporting score scale ζ, rather than doing a direct translation, panelists are asked to provide information through the use of a standard setting kernel that can be used to estimate the standard on the reporting score scale. This information is collected in the matrix $\mathbf{R}$ that typically has the information from each panelist in a row of the matrix.

As described in previous chapters, the information in $\mathbf{R}$ can take many different forms. It could be bookmark locations from an OIB, it could be a string of conditional probability estimates from the use of an Angoff kernel, it could be classifications of candidates as exceeding or not exceeding the requirements in $\mathbf{minD}_C$, etc. Then statistical/psychometric methodology is used to estimate the location of the cut score on the reporting score scale from the information in **R**.

Common assumptions of statistical estimation are that the observations used for estimation are independent and identically distributed (Lehmann & Casella, 1998). In the case of estimating a standard, that means the information provided by panelists is assumed to be independently obtained and that there is the same error distribution for each panelist. These assumptions support the computation of the usual statistics that are used to estimate the standard. In many cases, however, the information obtained from panelists is purposefully not independent. While the first-round implementation of a standard setting kernel is often done in a way that can be considered to give independent information, after the first round there is often discussion among panelists that likely bring about dependence among the information in the rows of **R**. From a statistical perspective, there is no impact on the mean of panelists' estimates of the standards, but there is an impact on the standard deviations of the panelists' estimates. Dependencies, such as the dominance of the discussion of initial results by one person, can reduce the standard deviation of estimates. That can be seen as a positive as moving toward consensus if the consensus is around the standard implied by $\mathbf{D_P}$. If the dependencies result in movement away from the intended standard, then it results in bias in the estimation process. This can happen if a panelist has an incorrect interpretation of $\mathbf{D_P}$ or does not accept that the task is to translate the policy proposed by the agency. If that panelist actively works to influence the other panelists, statistical bias in the estimate may be the result.

Depending on the standard setting kernel that is used, there can be an assumption of independence within the vector of information provided by panelists as well as between panelists. For example, when the Angoff probability-estimation kernel is used, panelists provide probabilities of correct response corresponding to persons with $\mathbf{minD_C}$ for each item. It is usually assumed that each of those judgments is made independent of the others. If an IRT model is used to define the reporting score scale for the test, each of those judgments can be used to get an item-specific estimate of the standard and the standard deviation of those estimates can be used as an estimate of the amount of error in the panelist's judgment. Dependence among those judgments will affect the estimate of the standard deviation.

There is one other assumption of independence that is typically made—independence of the work of different standard setting panels. Some standard setting experts (e.g., Clauser, Margolis & Clauser 2014; Lewis et al. 1998) organize panelists into separate groups so that they can serve as replications of the standard setting process. The variation in the results from these separate groups is used as an estimate of the amount of error in the process. The results from the different groups are assumed to be independent of each other. This would seem to be a reasonable assumption except that the groups often receive general information together from a single facilitator. This is done to make sure they have a common understanding of the process, but it also builds in some dependence. This dependence can be more or less contingent on how much the groups do in common. That information should be included in a detailed report of the standard setting process.

8.2.3 Other Assumptions

The two assumptions listed in the previous sections, the hypothetical continuum and independence of information provided by panelists, are augmented by other assumptions depending on the measurement model that is used to define the operational test score scale and the assumptions made by the standard setting kernel that is selected for the standard setting process. For example, if the measurement model is the Rasch model (Rasch, 1960), there are assumptions that the item-score data from the test can be well represented by a logistic function that contains a single person parameter and a single item difficulty parameter for each item. These assumptions imply others, such as the probability of correct response to a test item is monotonically increasing with an increase in the parameter representing the proficiency of examinees. This measurement model also implies that the discrimination power of test items can be represented by a single value that is common to all items and that it is not possible to obtain a correct response to a test item by guessing or some other process not related to proficiency.

There are other assumptions about panelists. They are assumed to be competent in the areas that are necessary for them to produce entries for **R** that are useful for estimating the standard. In the analogy to language translation, they are assumed to be proficient in both the initial language and the target language. The equivalent here is that they comprehend the information in the definitions and, after appropriate training, can perform the tasks required by the standard setting kernel. For example, if a Contrasting Groups kernel is used that requires the classification of students as competent or not, it is assumed that the panelists know the students sufficiently well to make the required judgments and that they will make the judgments about the target construct for the standard setting process rather than other characteristics of the person.

There are too many assumptions required by the many kernels and standard setting designs to list them all here. Specific assumptions will be listed when describing estimation for a particular standard setting process. It is important that reports of standard setting processes include summaries of the assumptions made when implementing the process.

8.3 The Value that Is Estimated

The question of what is being estimated from the information collected during a standard setting process seems obvious. In the theory proposed in this book, the thing that is estimated is the point on the hypothetical construct implied by the policy definition $\mathbf{D}_P$, which corresponds to the intended standard that is described by $\mathbf{D}_C$ and $\mathbf{minD}_C$. It is represented by ζ_{Ag}. But the construct represented by ζ is in the heads of the members of the agency who produce $\mathbf{D}_P$ and it is not observable to the panelists or those who design the standard setting process. It is only if the agency indicates that the recommended standard does not match the intentions implied by $\mathbf{D}_P$ that we know that there is slippage in the process that let it drift away from accurately capturing the information in all the definitions.

Panelists gain their understanding of the construct from $\mathbf{D}_P$ and $\mathbf{D}_C$, which by their nature are relatively brief summaries of the characteristics of the desired construct and the content domains. The panelists are also given experience with the test that was developed or selected as the operational definition of the intended construct. So, when panelists perform the tasks required by the kernel, are they thinking about ζ or some approximation of it? It seems likely that each panelist has developed their own conception of ζ that may not be exactly the same as that conceptualized by the agency. Here a panelist's construct is labeled $\zeta^{(p)}$ with the p referring to the specific panelist. In theory, each person in a population of interest has a location on both ζ and $\zeta^{(p)}$ and the population correlation between the locations is $\rho_{\zeta,\zeta(p)}$, that is 1.0 in the ideal case but is less if there is not complete understanding of the definitions by panelist p.

The construct ζ is conceptual and it does not automatically come with a scale associated with it. The construct *height* does not have an implicit scale associated with it, so different scales have been created depending on how the construct is being used: inches, centimeters, cubits, miles, etc. Eventually, conventions for describing height were developed and standardized. Most of the constructs considered in standard setting studies have not reached that level of sophistication and standardization. At best, the constructs underlying standard setting studies are kept constant over a number of years using equating methodology and they have a common test design to operationalize the scale.

To make these ideas more tangible, consider the following situation. An agency that is responsible for the education of the population of a municipality sets a policy that all those who complete compulsory education should be able to read English sufficiently well to function in society. The agency produces a policy statement that describes the kinds of texts that all citizens should be able to read and the level of competence that is expected, $\mathbf{D}_P$. Further, they convene a panel of reading experts to further describe the reading skills and knowledge required and the minimum level that would be accepted ($\mathbf{D}_C$ and $\mathbf{minD}_C$). Finally, the agency reviews and approves all these documents.

The example shows that the agency has a concept of what is meant by reading proficiency and the definitions that were produced had the purpose of communicating the characteristics of the reading construct to a wider audience. That construct is the continuum on which the standard for reading proficiency is defined. When describing this continuum and the standard, there is no mention of a scale of reading proficiency or of error associated with the standard. Conceptually, the standard is expressed like a sign at an amusement park that states "must be taller than this level to ride." There is an assumption that there is an accurate measure of the desired construct, and that the expression of the standard is precise.

In theory at least, it is possible to apply the policy defined by the agency without any scale for reading proficiency being defined. If a person has complete understanding of the definitions and a thorough understanding of the reading skills of persons who are to be evaluated relative to the standard, he or she could accurately classify persons as meeting or not meeting the standard. This is an application of the judgment task used with the Contrasting Groups kernel. Unfortunately, any description, no matter how detailed, is open to interpretation. It is likely that each person who reads the definitions will have slightly different interpretations of the nuances of the definitions. The result is that they may have a slightly different conception of the construct and the location of the standard on the construct. Further, conceptions of the construct and the location of the standard will likely differ slightly for each panelist. The practical consequence of this is that there will not be perfect agreement on the classification of persons as meeting the standard or not meeting the standard. In fact, there may be significant numbers of different classifications as shown by the research literature on scoring accuracy for essays.

Even though a continuum is implied for reading proficiency, no scale has been defined for the continuum. There is an implication that individuals differ along the continuum, but the only descriptive information is about the meaning of the construct and the point on it that corresponds to the standard. The Contrasting Groups kernel can still be applied in this situation. Panelists can be asked to classify individuals as meeting or exceeding the standard, or of not meeting the standard. However, many assumptions are needed to be able to tell whether a panelist is accurately translating the definitions into appropriate applications of the standard. The only data that are available are the classifications of individuals and perhaps a description of the group that was classified. It may be that each panelist has defined the construct in a unique way and has a unique interpretation of the meaning of the definitions.

Because the true classifications of individuals relative to ζ and ζ_{Ag} are not known, it is not possible to tell if panelists are accurately making the intended classifications. It is possible to gather information about whether panelists are classifying individuals in the same way if they classify the same individuals, and they have the same level of familiarity with those persons. Insights into the results that can be expected from analyses of classifications of individuals can be gained from the research literature on scoring of essays. For example, Hollenbeck, Tindal, and Almond (1999) report that for the evaluation of six distinct characteristics of an essay, there was on average 24.4% of the essays with disagreements about whether they had exceeded the state standard for writing performance. This result was from persons trained to evaluate the essays using a common set of criteria. Comparable results might be expected if trained panelists are making classifications of individuals relative to ζ_{Ag}.

There are several ways to explain the differences in classifications, even if all the classifications are based on the true scores for the individuals. The panelists may have different conceptualizations of $\mathbf{minD_C}$ and the corresponding location for ζ_{Ag}, they may have a different conceptualization of ζ, or have different conceptualizations of both. Even though the panelists may have accurate assessments of the skills and knowledge of the persons being classified, they may be making those classifications using somewhat different criteria. Because the designers of a standard setting process do not know which panelists have accurate conceptions of ζ and ζ_{Ag} and which do not, the hope is that the differences in interpretations will be offset, so an aggregated result will be close to the intentions of the agency. For example, 15 panelists might be asked to classify each individual relative to the standard and the most frequent classification could be used as an estimate of that intended by the agency.

In real situations, there are few cases where multiple panelists have experience with the same persons; so the hypothetical process in the example in the previous paragraph is not practical. Instead, a measurement device is developed or selected to serve as an operational definition for ζ. Then the information obtained through the application of the standard setting kernel is used to estimate the location of ζ_{Ag} on the continuum defined by the test.

The real estimation problem for estimating ζ_{Ag} is more challenging than described so far because although panelists may conceptualize the continuum and point on the continuum corresponding to $\mathbf{minD_C}$ without error, their judgments contain error and the scores on the reporting score scale for the test that is the approximation of ζ also contain error. Even when the Contrasting Groups kernel is used, the classification of persons by panelists is based on their estimates of locations on the continuum. The literature on the scoring of open-ended items by trained readers indicates that the reliability of those estimates is not high. For example, Cohen (2017) reported a range of reliabilities for raters from .47 to .70. The implication is that the information from the application of the Contrasting Groups kernel contains variation due to different understandings of $\mathbf{minD_C}$ and ζ and the evaluation of candidates by panelists is not very reliable. The goal of the design of a standard setting process is to acquire information from panelists that minimizes variation in understandings of the intentions of the agency and maximizes the reliability of the ratings obtained from panelists. The result is the information contained in $\mathbf{R}$. Given that information, the goal is to get an estimate of the reporting score scale that is an accurate representation of the mapping of ζ_{Ag} to the reporting score scale.

8.3.1 Estimation Based on Contrasting Groups Data Collection

To demonstrate the impact of all these features of the standard setting process, an example is provided based on the Contrasting Groups kernel. Suppose that, as described earlier, a standard is being set for proficiency in reading English text as part of a goal for compulsory education in a school system. It is assumed that performance on the construct defined by the $\mathbf{D_P}$ can be described by the standard normal distribution and that the intended standard is such that 84% of those being evaluated meet the standard based on "true" scores. That is, the intended standard ζ_{Ag} is at –1 on the scale implied by the standard normal distribution.

Suppose also that 15 experts on the language curriculum and the examinee population have been selected for the standard setting process because they have extensive personal knowledge of a sample of 100 persons from the population. These 15 experts are given the $\mathbf{D_P}$, $\mathbf{D_C}$, and $\mathbf{minD_C}$ and they discuss those definitions with some members of the agency to get a good conception of the intended construct and the intended standard. However, they each have their own nuanced interpretation of ζ and ζ_{Ag}. In the example, a panelist's understandings of the construct $\zeta^{(p)}$ are correlated above .80 with ζ and their understanding of ζ_{Ag} is within .2 of the intended standard on the scale defined by the standard normal distribution.

To see the influence of the panelists' interpretation of the construct and the standard, and the accuracy of the estimates of candidate performance on the estimation of the cut score for contrasting groups, two examples are provided below. One is for a panelist with a good understanding of the construct. The correlation between the true measures on the construct and the panelist's (Panelist 4) conception is .99, but the panelist's understanding of the standard is –1.09, somewhat lower than the intended standard of –1. This panelist also has reliability of judgments of .53, a moderate value for this type of judgment. Figure 8.1 presents scatter plots of the panelist's estimate of the location of candidates' positions on the construct against the true locations of the candidates on the construct defined by $\mathbf{D_P}$. The figure also includes the intended cut score on the horizontal axis and the panelist's interpretation of it on the vertical axis.

Figure 8.1 contains two scatter plots. One is a plot (a) of the true score for Panelist 4's interpretation of the construct against the true score on the construct intended by the agency. The other is a plot (b) of the panelists' scores with estimation error on the interpretation of the construct against the true score on the construct intended by the agency. In the first case without error, there are only 3 candidates out of 100 that are classified differently based on the panelist's interpretation of the construct and standard. These were all classified as above the standard because Panelist 4 interpreted $\mathbf{minD_C}$ as slightly lower than was intended by the agency.

When error is included in the panelist's judgments about candidates, there is an increased spread of locations for the candidates and there is a much greater number of differences in classifications of the candidates above and below the standard. There are now 18 candidates that have different classifications—13 classified below the standard by the panelist when they were above the standard intended by the agency. The results of the interpretation of the construct and the standard by Panelist 4 is that the proportion of candidates estimated as above the standard is different than that intended by

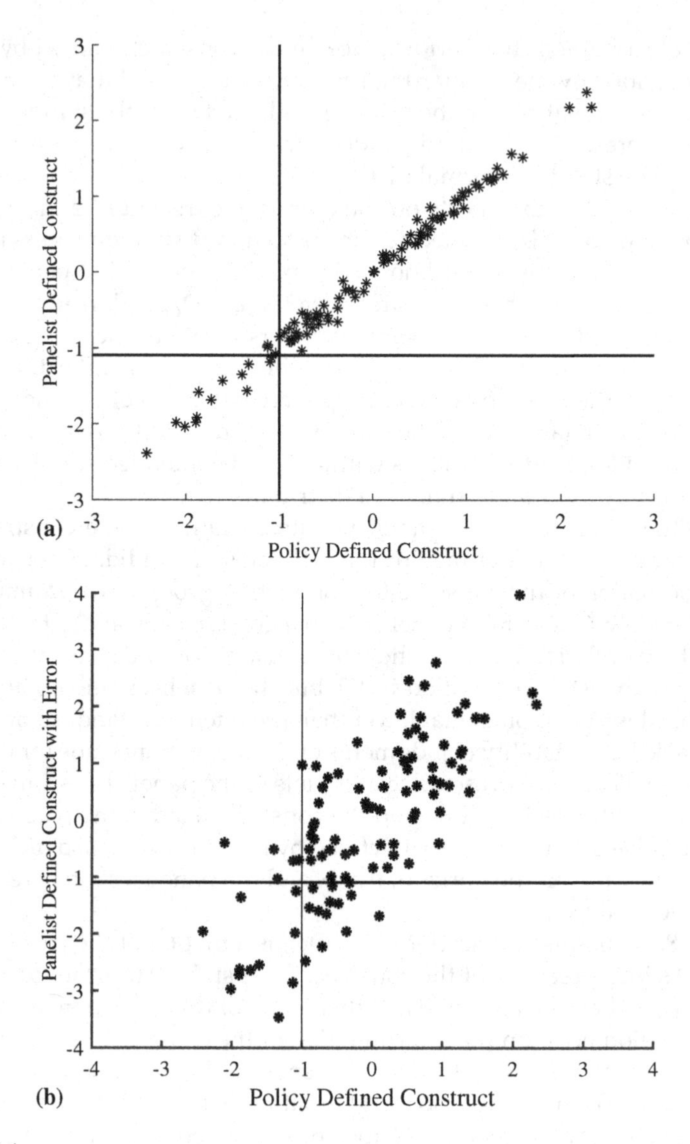

FIGURE 8.1
Scatter plots of locations of candidates comparing true scores on the construct defined by the policy definition and the interpretation by Panelist 3 without (a) and with (b) error estimating candidates' locations.

the agency (84%)—87% when there is no error in estimating the interpreted construct and 76% when judgment error is included. These results show that this panelist interprets the standard as being lower than intended resulting in a higher proportion passing. Error in judgments spreads out the estimates of locations of candidates on the scale resulting in fewer than intended being classified as above the standard.

A second panelist, #2, has interpreted the construct differently than Panelist 4—the correlation between locations on Panelist 2's construct and the agency's intended construct is .81 rather than .99 for Panelist 4. Panelist 2 also interprets the standard as being slightly lower than the intentions of the agency: –1.0624 rather than –1. Figure 8.2 presents the scatter plots for

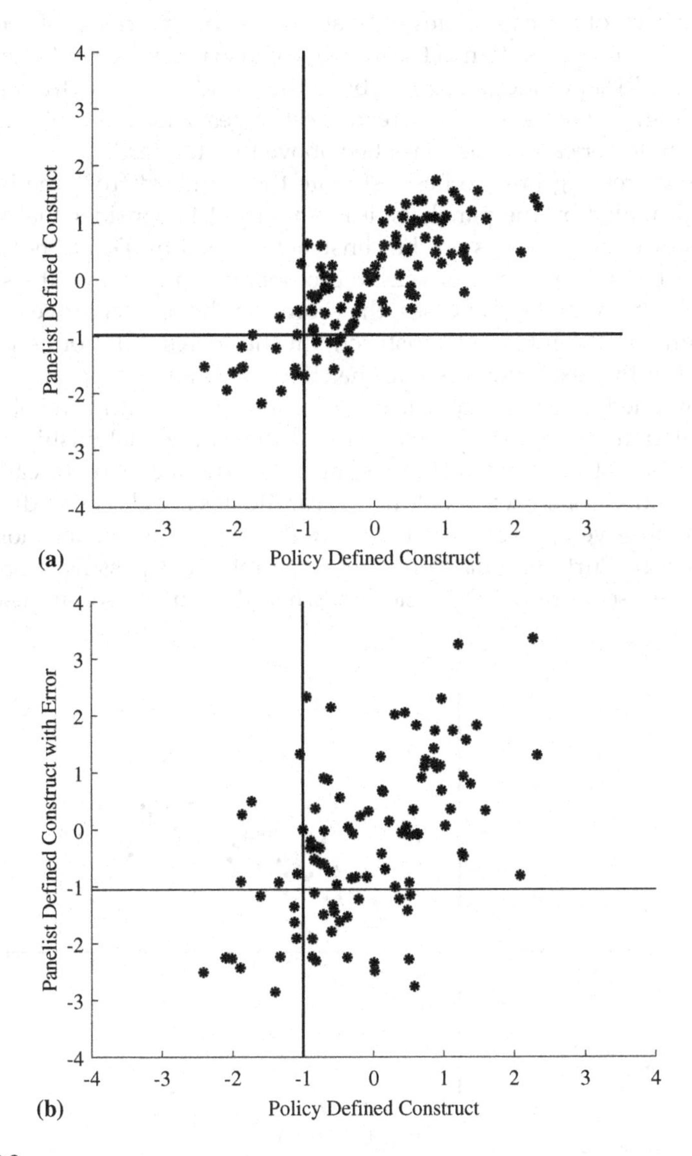

FIGURE 8.2
Scatter plots of locations of candidates comparing true scores on the construct defined by the policy definition and the interpretation by Panelist 2 without (a) and with (b) error estimating candidates' locations.

Panelist 2 construct against the intended construct of the agency, again without (a) and with errors (b) of judgment.

It is clear that the change in interpretation of the construct increases the spread of locations in the scatter plots and that the lower interpretation of the standard results in more candidates above the standard than was the case for Panelist 4. When there was no error in the classifications of candidates by Panelist 2, 82% of the candidates were above the interpretation of $\mathbf{minD_C}$ on that panelist's interpretation of the intended construct. When judgment error is included, 71% are judged to be above the standard. As with Panelist 4, error in judgment increases the spread of estimates of locations of candidates resulting in fewer candidates classified above the standard.

Before addressing how to best estimate the standard from the information, **R**, provided by the panelists, it is important to consider that the true locations of the candidates on the construct defined by $\mathbf{D_P}$ are not known. Instead, a test is developed or selected to operationally define the scale for the construct. In the best of cases, the scores on the test will give estimates of locations on the intended construct, but the scores will not be perfectly reliable. For this example, assume that the test is an appropriate measure of the intended construct, and it is well constructed with a reliability for the population of interest of .90. Figure 8.3 shows the relationship between the observed estimates from the test and the true locations of candidates on the intended construct. Six of the 100 candidates are classified differently using the observed scores from the test rather than the true locations from the construct. Further, 80% were above the intended passing score using the observed score rather than the 84% when the true scores are used. The

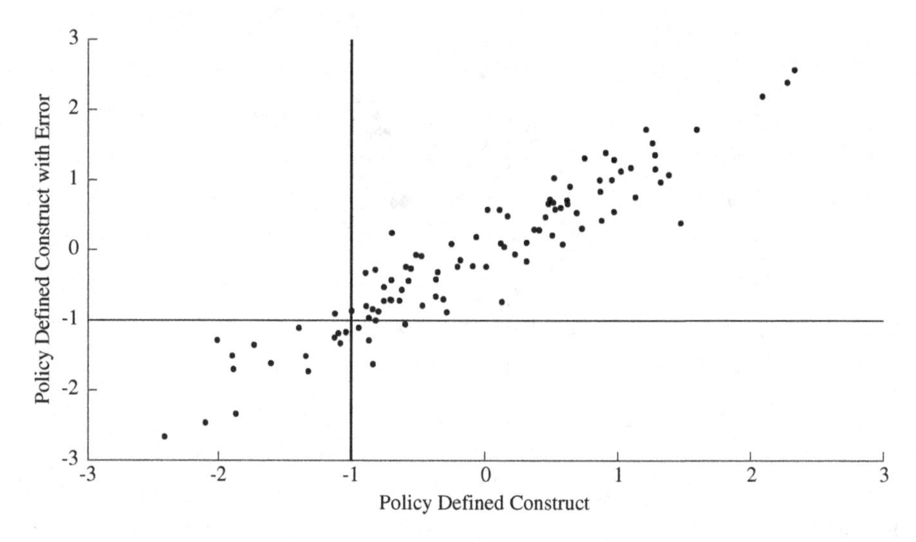

FIGURE 8.3
Scatter plot of locations of candidates comparing true scores on the construct defined by the $\mathbf{D_P}$ and observed scores from the test used to operationalize the construct.

percentage for the observed scores is lower because adding error to the scores increases the standard deviation of the distribution and, of course, there is also sampling variation because the sample size for these statistics was 100.

When the agency and the panelists are thinking about standards on the intended construct, ζ, it seems unlikely that they are thinking about the observed scores. The definitions are abstract, and they do not typically contain any reference to error of measurement or operational definitions. Further, when classifying candidates, they are not told to account for errors in classifications. This implies that the goal of estimation in standard setting is to translate $\mathbf{D}_\mathbf{P}$ to a point on the true construct rather than on the observed score scale for the construct. In a practical sense for this example, it means that the true percentage to be estimated is 84% and the true value of ζ_{Ag} is -1 on the scale for the assumed normal distribution of the population of candidates. The goal for estimation within the standard setting process is to use the information in **R** to best approximate these true values implied by the definition.

The standard setting process is essentially a data collection design for accumulating information from panelists that is related to the location on the construct defined by $\mathbf{D}_\mathbf{P}$ that corresponds to the standard intended by the agency. In the example provided above, there is a particular data collection design based on certain assumptions and design considerations. These assumptions and design considerations are as follows:

1. There are 15 panelists who are selected because they are knowledgeable about the construct and the candidate population.
2. The panelists have all been trained on the context for the standard and $\mathbf{D}_\mathbf{P}$, $\mathbf{D}_\mathbf{C}$, and $\mathbf{minD}_\mathbf{C}$. However, it is assumed that they have slightly different interpretations of the construct and the location of the standard on the construct.
3. The panelists evaluate each of 100 candidates that have been randomly sampled from the population of candidates. The panelists indicate the candidates from the sample that they judge to be above their interpretation of the standard.
4. The panelists have enough familiarity with the sample of candidates to make reasonable judgments about their capabilities relative to the standard, but some panelists are more accurate at making these judgments than others.
5. The information in **R** is the classification of each candidate (1 for above the standard and 0 for below the standard) by each panelist. It is a 100 × 15 matrix of classifications with rows as candidates and columns as panelists.

Table 8.1 summarizes the information about the panelists' understanding of their task, including their interpretation of the construct (correlation between

TABLE 8.1

Information about Panelists Interpretation of the Standard Setting Task

	Interpretation of Scale and Standard			Data for Estimating Standard		
Panelist	Correlation with ζ_{Ag}	Interpretation of Standard[a]	Reliability of Judgments	Percent above Standard[b]	z-score Based on Standard Normal Distribution	z-score Based on Observed Score Distribution
1	.99	–1.10	.54	77	–.74	–1.00
2	.82	–1.08	.53	71	–.55	–.77
3	.99	–0.93	.63	82	–.92	–1.11
4	.98	–1.01	.41	76	–.71	–.97
5	.84	–0.98	.52	76	–.71	–1.04
6	.86	–0.97	.50	76	–.71	–.87
7	.96	–0.97	.59	73	–.61	–.81
8	.90	–0.91	.41	74	–.64	–.86
9	.89	–1.02	.43	70	–.52	–.68
10	.85	–1.09	.68	73	–.61	–.96
11	.93	–0.96	.59	71	–.55	–.76
12	.96	–0.95	.43	74	–.64	–.87
13	.94	–1.00	.62	71	–.55	–.73
14	.82	–1.09	.46	67	–.44	–.63
15	.86	–1.07	.59	73	–.61	–.82
Mean	.91	–1.01	.53	73.6	–.63	–.86
Median	.90	–1.00	.53	73	–.61	–.86

[a] Intended standard is –1.00. [b]True percent above agency intended standard is 84.

their construct and that of the agency), interpretation of the standard (location of their standard on their construct), and reliability of their judgments. The last two columns of the table report some initial results of attempts at estimating the intended standard. The information in those columns will be discussed after highlighting some information about the quality of the simulated panelists' interpretations of construct and standard.

First, the correlations between the panelists' interpretation of the construct and the intended construct are generally high. Some are .99 indicating that panelists are modeled as having a very good understanding of the intended construct. The lowest correlation is .81 which seems high as well, but the scatterplot for that panelist is given in Figure 8.2 and it shows noticeable spread in this simulated panelist's locations for candidates with the same location on the intended standard. The intention here was to model the influence of different interpretations of $\mathbf{D_P}$ and $\mathbf{D_C}$ by panelists. The results are similar for the modeled interpretation of the intended standard of –1 on the intended construct. Some of the simulated panelists have interpretations that are very close to that value, but others are somewhat lower or higher than the intended standard. For example, Panelist 1 has an interpretation of –1.10 which is lower than the intended standard. Panelist 8 has a modeled interpretation of –.91 which is harder than the intended standard.

The reliability of judgments for the modeled panelists is relatively low compared to what is typical for test scores. These values are based on the literature on reliability of scoring essays. The reliability of a trained essay scorer is in the range shown here. The expectation was that single panelists estimating the location of a candidate would have a reliability that was similar to what is observed for essay scoring.

Table 8.1 provides the results for this standard setting process in the last two columns. The only observable data from this standard setting design is the classification of the 100 sampled candidates as above or below panelists' interpretations of the standard. The number observed to be above the standard is given in the fifth column of the table. The result is that the simulated panelists judged fewer candidates as above the standard than was intended by the agency (mean of 73.6 compared to 84). The last column of the table shows the estimated locations of the standard on the scale of the standard normal distribution. These values were obtained by applying the inverse normal transformation to proportions of candidates judged to be below the standard by each of the simulated panelists. The results show how far the estimated standards are on average from the standard intended by the agency. The mean estimated standard is –.63 rather than the –1 intended by the agency. An important question is why that is the case, and would a similar result be expected in an actual standard setting study? Also, is it possible to identify a better way of estimating the standard that comes closer to reflecting the intentions of the agency?

The explanation for the result is that the effect of the modeled judgment error and differences in interpretation of the construct is to increase the variance of the estimated locations of the candidates. Yet standard setting studies assume that panelists have a good understanding of the intended construct and the intended standard. In this simulation, those are treated as the equivalent of true scores. Because the estimated locations of candidates have more variation than the true scores on the intended construct, more candidates are estimated to be below the standard. This result is specific to standards that are below the mean. If the standard had been very high, for example, to select scholarship candidates, the opposite effect would be observed. There would be more candidates estimated to be above the standard.

For this example, the variance of the true values on the intended construct for the sample of 100 candidates was 1.02. The population standard deviation is 1.0 and the sample value is close to that of the full population resulting in 84 candidates above –1 as expected. The average variance for the locations from the panelists including their estimation error for the 100 candidates is 1.95. This is very close to what would be expected based on the reliability of the judgments. If the true score variance is 1.00, the estimated reliability based on the average variance from the observed locations is 1.00/1.95 = .51, similar to the observed average reliability of .53.

Under the conditions of this simulation, the estimated standard is notably higher than the standard intended by the agency. The opposite finding would be expected for intended standards above the mean of the distribution of candidate locations. There is a regression effect for the estimation of the location of the standard because of the influence of errors in the judgments of locations of the candidates by the panelists. The question to be addressed here is how best to estimate the location of the standard on the intended construct to adjust for the observed regression effect.

In the case where the reliability of panelists' judgments is known, the value of the reliability can be used to adjust the estimate of the standards. For example, the reliability for Panelist 1 was modeled as .54 in the example. If the true variance of the intended construct is set to 1.0, the observed variance is $\sigma_o^2 = \sigma_t^2 / r_{xx}$, where r_{xx} is the reliability for the panelist. For this case, the observed variance would be 1/.54 = 1.85. The initial estimate of the standard was obtained using the inverse normal transformation of the proportion above the standard assuming a standard normal distribution. That resulted in the value of –.74 for the estimated standard when the standard intended by the agency was –1. However, if a normal distribution is assumed with a mean of 0 and a variance of 1.85, when the inverse normal transformation is performed, the estimated standard is –1.00.

Looking at one panelist and correcting for reliability would seem to solve the estimation problem. However, the situation is more complex. The last column of Table 8.1 contains the estimated cut scores for each panelist when the reliability of judgments was taken into account. The mean estimate over the 15 panelists was –.86 rather than the –1 intended by the agency. This would seem discouraging, but because this is a simulation, it can be replicated numerous times. The mean of the mean-estimated-standard over ten replications was –1.01. The range of estimates was from –.86 to –1.21. This indicates that with either more panelists, or more individuals to classify, a better estimate of the standard could be obtained.

There is the temptation here to investigate the sample size requirements of this implementation of the Contrasting Groups kernel. However, this implementation of Contrasting Groups is not widely used. It is a practical standard setting kernel if panelists are classifying examples of work, such as essays or artwork, allowing all the panelists to classify the same set of examples. However, the kernel is based on the assumption that sample of work examples is representative of the population so that the proportion classified above the standard by each panelist is an estimate of the proportion in the entire population. It is the rare case that this assumption can be met. Also, even if a random sample of examples of work was used, the application of the estimated standard to other examples of work requires the same set of panelists to do the classifications. It is unlikely that a group of panelists can classify thousands of papers or work samples. The usual approach is to use the data from the classifications of papers or examinees to estimate the standard on the reporting score scale for the test that operationally defines the construct. Then, decisions about quality of the work samples or examinee performance can be made by comparing scores for those samples to the cut score on the reporting score scale. Two approaches for estimating the score on the reporting score scale are discussed below.

For the simulation presented above, an observed score for the intended construct based on a reliability of .90 was also generated. Figure 8.3 presents a scatter plot of the observed scores against the true scores for candidates. To determine how to estimate the standard on the observed score scale from the panelists' classifications, it is useful to know the score on that scale that best matches the standard intended by the agency, –1. One way to decide on the corresponding cut score on the observed score scale is to find the score that maximizes the proportion of correct classifications for the examinees. The graph in Figure 8.4 shows the proportion of correct classifications for possible cut scores on the observed score scale based on a random sample of 1,000 candidates. The maximum proportion is .95 using a cut score of –1.1 so that value will be used as the target for estimation for the information obtained from the panelists.

The proportion of correct classifications used here assumes that false positive classifications and false negative classifications have equal importance. A different value for the cut score would be obtained if errors in classification above or below the standard were given different weights.

A common statistical approach to identifying the cut score on the observed score scale is discriminant analysis. There are many ways to apply discriminant analysis to this problem because it allows weighting of false positive and false negative classifications and many other options. The typical application also assumes normal distributions of the variable of interest for each

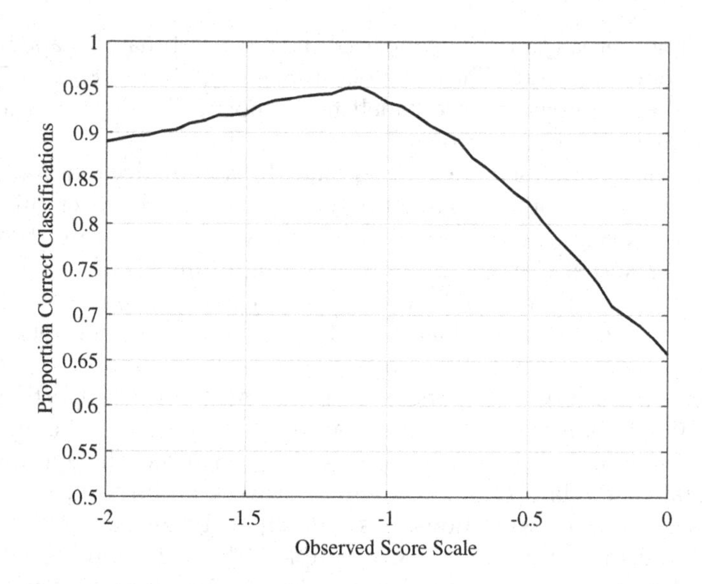

FIGURE 8.4
Proportion of correct classifications for possible cut scores on the observed score scale compared to a cut score of –1 on the true score scale.

classification group. Applying discriminant analysis to the observed scores on the hypothetical test with the true classifications of the candidates yields a cut score of –1.176 for minimizing the classification error for the candidates.

Another statistical approach to estimating the cut score is logistic regression (e.g., Plake, 1998). This approach uses the pass/fail classification as the dependent variable and the observed test score as the independent variable. Applied to the observed scores on the intended construct and the true classifications, logistic regression yields an estimate of –1.165. Other standard setting experts (e.g., Hambleton, 1998) suggest using the location of the intersection of the distributions of the observed scores for the groups of passing candidates and failing candidates. Figure 8.5 shows smoothed frequency distributions of the observed scores for those classified as passing or failing on the intended construct. Two smoothing methods were used. One was a non-parametric approach called kernel smoothing. This method makes no assumptions about the shape of the score distribution and shows some of the irregularities in the shape. The other method is to fit a normal distribution to the observed score distribution. The cut score on the observed score scale was determined by where the pairs of distributions intersect yielding –1.25 for the kernel smoothing and –1.27 for the normal distribution estimates. Table 8.2 presents the results for all the different approaches to estimating the cut score for convenient comparison.

The first method in Table 8.2 is simply finding the cut score that would have the same proportion above it as in the sample of 1,000 candidates used for this simulation. It does not consider error in classification resulting from the error in the observed test scores. That was the highest cut score and the

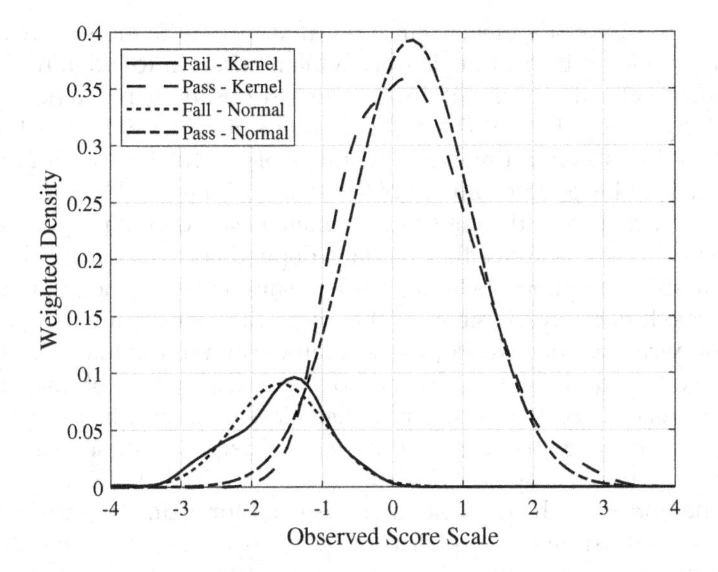

FIGURE 8.5
Smoothed distributions of observed scores for candidates with true classifications of passing or failing.

TABLE 8.2

Cut Scores on the Observed Score Scale Corresponding to –1 on the Intended Construct

Method	Cut Score	Percent Passing
Same as true percent passing	–1.075	85
Correction using reliability	–1.139	87
Maximize correct classification	–1.1	86
Discriminant analysis	–1.176	87
Logistic regression	–1.165	87
Intersection of distributions (kernel smoothing)	–1.25	89
Intersection of distributions (normal distributions)	–1.27	89

lowest percent passing. All the other methods take the error in the test into account in different ways resulting in slightly different values for the cut scores. All result in a slight positive bias in the percent exceeding the cut score from 1% to 4%. That bias in the percent passing is a result of accounting for errors in classification. The bias would be negative for cut scores above the mode of the observed score distribution. The largest bias occurs when the intersection between the two distributions is used to set the cut score. That is the point where the probability of being in each of the distributions is equal when the proportions passing and failing are taken into account. This result merits further research to determine if it is a consistent finding. It may be a reason for not using this method.

Most of the methods yield a cut score that rounds to either –1.1 or –1.2. The purpose for this example is to provide a criterion to judge the quality of estimates of cut scores obtained from various implementations of the Contrasting Groups kernel. The observed range of values will serve that purpose. If methods yield estimates in the range of –1.1 to –1.2, they will be considered to provide good estimates of the standard intended by the agency.

As noted previously, the cut score estimate based on the proportion classified by panelists as above the standard depends on the samples of candidates classified by panelists being random samples from the population of candidates. It is rarely possible for that to be the case even when panelists are classifying papers or work products rather than individuals. A common alternative approach is for panelists to classify a sample of convenience, get the observed score for each candidate classified, and then estimate the cut score from the distributions of those classified as being above or below the standard implied by $\mathbf{D}_C$. The data from Panelist 1 in Table 8.1 is used here to demonstrate the possible methods used for estimating the standard using the Contrasting Groups kernel. Figure 8.6 shows the normal distributions with the same mean and standard deviations as the observed scores for the candidates classified as above or below Panelist 1's interpretation of the definitions of the standard. There is substantial overlap in the distributions because of the error in judging candidates and the error in the observed scores. Table 8.3 provides summary statistics for the distributions.

For this set of 100 observations for simulated Panelist 1, the discriminant analysis provides an estimate of the cut score that is closest to the target determined in the analysis given above. The other estimates of the cut score

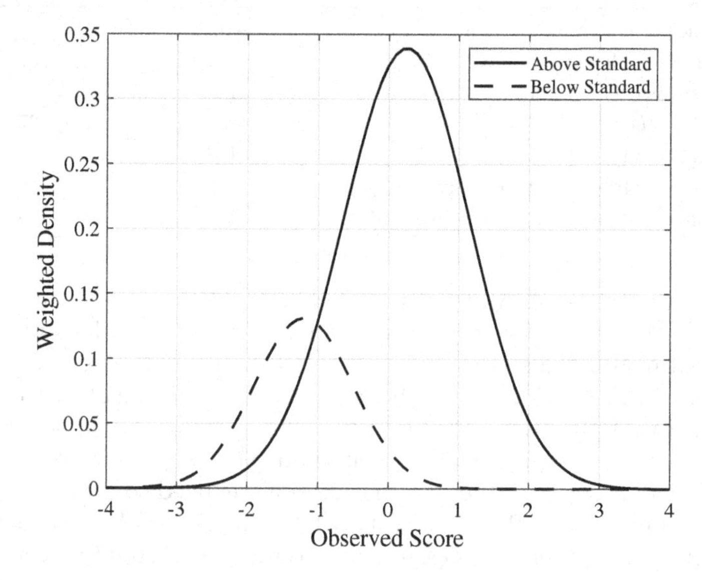

FIGURE 8.6
Normal approximations to observed scores for candidates classified by Panelist 1.

TABLE 8.3

Summary Statistics and Cut Score Estimates for Panelist 1

Candidate Classification	Mean	Standard Deviation	Method		
			Intersection	**Logistic**	**Discriminant**
Above	.25	.90	–1.02 (83)	–1.04 (84)	–1.09 (85)
Below	–1.20	.70			

Note: Values in parentheses are percent above the cut scores.

are somewhat higher resulting in slightly lower percent of candidates above the cut score. Because these results are based on a simulation of the judgments and the error in the observed scores, the estimation process can be replicated to determine how much variation there is in the estimates. To check the accuracy of the estimates, the cut score was estimated using each method using ten independent samples of 100 simulated candidates. Table 8.4 presents the results.

The results in Table 8.4 show that all three methods gave approximately the same estimate of the cut score for the situation when one panelist with a good understanding of the intended construct (Panelist 1) and the intended standard classifies 100 work samples with a reliability of .54. All three methods estimated the cut score as approximately –1.2 and all three had about the same amount of variation in the estimates. This is a positive result for this method of estimating the standard using the Contrasting Groups kernel. However, this result applies to the single panelist with the defined characteristics. A more interesting question is whether the estimation procedure will work equally well when 15 panelists with varying characteristics classify the same set of work samples?

Table 8.5 presents results that are similar to those that might be obtained from an actual standard setting process where 15 panelists applied the Contrasting Groups kernel to the same set of 100 work samples such as essays, performance tasks, etc. Those work samples had been independently scored and the panelists classified the work samples according to their

TABLE 8.4

Mean of Estimated Cut Scores for Ten Replications of the Three Methods for Panelist 1

Statistic	Method		
	Intersection	**Logistic Regression**	**Discriminant Analysis**
Mean	–1.22	–1.19	–1.23
Standard Deviation	.23	.22	.21

TABLE 8.5

Descriptive Statistics for the Estimated Cut Scores for 15 Panelists Using the Three Methods on the Same 100 Work Samples

	Method		
Statistic	**Intersection**	**Logistic Regression**	**Discriminant Analysis**
Mean	–1.09	–1.15	–1.18
Median	–1.08	–1.11	–1.15
Standard Deviation	.21	.22	.24

Note: Standard deviations were computed on the estimates for the 15 panelists.

interpretation of the definitions and construct intended by the agency. Table 8.5 gives the results for the same set of work samples classified by Panelist 1 with results shown in Table 8.3. Note that by chance, the set of work samples tended to give somewhat lower estimates of the standard than was shown by the replications in Table 8.4. Variation in the work samples and differences in classification error clearly have an influence on the results. Table 8.4 also shows the median of the estimates from the 15 panelists because that statistic is often used to reduce the influence of outliers on the aggregated estimate of the standard. In this case, the medians are all lower than the mean values suggesting that there may be some extremely high estimates of the standard from some panelists that influence the mean estimates.

The influence of the sources of variation can be observed from the results obtained from replicating the full standard setting process with the 15 panelists ten times. Table 8.6 presents those results. The results are similar to the results for Panelist 1 with ten replications. The estimate of the standard is approximately 1.2–1.3 with the slightly higher value for the discriminant analysis. The standard deviation of the estimates is smaller than when a single panelist was considered: .11–.13 versus .21–.23. This reduction was expected because much more data was being used to obtain the estimates.

TABLE 8.6

Means and Standard Deviations of Estimated Cut Scores for Ten Replications of the Standard Setting Process with 15 Panelists Using the Three Methods

	Method		
Statistic	**Intersection**	**Logistic Regression**	**Discriminant Analysis**
Mean	–1.22	–1.23	–1.27
Standard deviation	.11	.13	.13

Note: Standard deviations were computed over replications.

The results for this particular implication of the Contrasting Groups kernel—15 panelists categorizing the same set of 100 work samples—are both encouraging and somewhat discouraging. It is encouraging that the replicated results are very close to the standard intended by the agency (see Table 8.2). The results are somewhat discouraging because the estimated standards from a single set of 100 work samples show the variation that can result from the sources of error in the judgment process. The differences in the single sample and replicated results are small, but these results are based on simulated data. Real data are always more variable because simulations cannot possibly include all sources of variation.

The Contrasting Groups kernel is applied in many different ways. It is useful for the discussion of estimation of standards to consider one other approach to the use of this kernel. It is often used to classify individual persons, students, or employees, and then using the test scores for those persons and the statistical approaches described here to estimate the standard. The difference of this approach and that described above is that the panelists typically only have sufficient knowledge of a small group of persons, like students in a teacher's class, and panelists are not classifying the same group of individuals. To compare this situation to the simulation described above. Consider the following scenario: Fifteen teachers each classify 50 students from their classes. Each teacher classifies a different group of 50 students. That means 15 × 50 = 750 students are classified in all, but only one teacher (i.e., panelist) classifies each student. Then the test results are collected for the students, and the classifications and scores constitute the information in **R** for estimating the point on the reporting score scale corresponding to the standard.

As with the previous example, the estimates of cut scores were computed using the three statistical procedures using the data from each panelist along with the standard deviation of the estimates over panelists. Also, the data from the 15 panelists were aggregated together and an estimate of the cut score was obtained from the full set of data. Table 8.7 contains these results.

TABLE 8.7

Estimates of the Cut Score Using Classifications of 50 Unique Individuals Per Panelist and the Aggregation of All Panelist's Classifications Using the Three Statistical Methods

	Estimation Method		
Summary Method	**Intersection**	**Logistic Regression**	**Discriminant Analysis**
Mean	–1.14	–1.45	–1.36
Standard deviation	.36	.77	.82
Median	–1.13	–1.09	–1.15
Aggregate data	–1.22	–1.21	–1.22

The results show some difficulties with the estimation of cut scores for the individual panelists. The sample of 50 students per panelist sometimes results in very few classified below the standard. This results in instabilities in the logistic regression and discriminant analysis that yields extremely low estimates for the cut score. Those few extreme estimates influence the mean of the estimates much more than the medians. This is a reason the median is often used for estimating the standard—that statistic ignores the distance of values from the center of the distribution. It is interesting that the intersection between the distributions of observed scores for those classified above the standard and those classified below the standard seems more robust than the more sophisticated statistical methods. This may be a result of stabilizing effect of approximating the data with normal distributions. Clearly, more research needs to be done on this topic. The review of the literature done for his book did not identify any research comparing methods for estimating the cut score using the Contrasting Groups kernel.

The aggregation of all the data gave essentially the same estimate of the standard for the three methods, about –1.22. This is a lower value than the medians reported for the individual panelists' estimates. The aggregated values are based on one sample of 750 cases. Although this is a relatively large number, it is useful to get an indication of how much variation there would be in the estimates if the 15 panelists replicated the judgments on multiple samples of students. As with the other examples, ten samples were used for the replication. Table 8.8 contains the results. These results show that the standard deviations are consistently small relative to the other ways of estimating the cut score and the mean standards are lower. The single aggregate results shown in Table 8.7 are higher estimates than those shown for the mean replications. This shows the possible problem with basing the standard on a single sample. However, the results reported in Table 8.8 yield 89% of candidates above the standard compared to the true value of 85% built into the simulation. The varied interpretations of the construct and the location of the standard, and the rating error of the panelists, result in this low estimate of the standard. It is important to understand why this occurs, so more detailed analysis of the sources of error in the standard setting process is provided.

TABLE 8.8

Means and Standard Deviations of Estimated Cut Scores for Ten Replications of the Aggregated Standard Setting Process ($N = 750$) with 15 Panelists Using the Three Methods

	Method		
Statistic	**Intersection**	**Logistic Regression**	**Discriminant Analysis**
Mean	–1.28	–1.27	–1.29
Standard deviation	.07	.07	.07

Note: Standard deviations were computed over replications.

A standard setting process seems to be estimating a point on a single score scale—the observed reporting score scale for the test that has been administered to candidates. But, from a theoretical perspective, there are really four types of score scales behind the standard setting process. The simulation of the standard setting process using the Contrasting Groups kernel used these four scales for generating data and analyzing the results. Table 8.9 lists these scales with descriptive statistics for each that show the influence of error on the basis for the standard setting process.

Two of the four scales listed in Table 8.9 are theoretical in the same sense as true scores in classical test theory. The agency true-score scale is the theoretical representation of the continuum defined in $\mathbf{D}_P$. The panelists have their own interpretation of the theoretical construct. In this case, for the aggregated 750 examinees, the correlation is .89 between the two versions of the true score scale. This is slightly lower than the mean value given in Table 8.1 because the data for the 750 examinees is a mix of the interpretations from the 15 panelists adding some additional noise to the data. The observed score scale for the panelists is not observed in practice because they are only asked if the examinee is above or below the standard, not the specific value that they used for making that judgment. The panelists' observed score scale contains the error in judgments plus the difference in interpretation of the

TABLE 8.9

Descriptive Statistics for the Four Scales Used When Generating the Simulated Data for Standard Setting Panelists Rating Unique Samples of 50 Examinees

	Scale Definition			
Statistic	**Agency True Construct Scale**	**Agency Observed Scale**	**Panelists' Interpretation of Agency Scale**	**Panelist Observed Scale**
Mean	.0074	−.0077	.0066	−.0792
Standard deviation	1.0166	1.0613	1.0099	1.3528
Proportion above standard	.8573	.8347	.8440	.7533 (.88)
		Correlations		
Agency true		.94	.89	.60
Agency observed	.9463		.85	.56
Panelists' interpretation	.8930	.8357		.77
Panelists' observed	.6410	.5924	.7360	

Notes: The only data that are observable during a standard setting study are the shaded cells. The value in parentheses is the proportion above the estimated standard using data from the Agency Observed Scale. Correlations below the diagonal are Pearson Product Moment correlations between locations of examinees on the score scales. Correlations above the diagonal are tetrachoric correlations between classifications relative to the cut scores.

agency's construct. It has a fairly low correlation, .59, with the observed score on the test that operationalizes the agency's construct. Note the large standard deviation of the observed scores for the panelists. This is a result of the errors in judgment, and errors in interpretation of the construct and standard.

The theory presented here includes all these scales in the generation of the data used for estimating the standards. However, in the typical standard setting study using the Contrasting Groups kernel, only the information in the shaded cells in Table 8.9 are available for use—the observed scores on the test selected by the agency and the classifications of examinees made by panelists. For this example, panelists estimated about 75% of the examinees were above the standard. Using those classifications and the observed scores for the examinees, the standards given in Table 8.7 in the row labeled "Aggregate data" were obtained. Using those estimated standards and the observed scores for the 750 examinees, about 88% of the examinees (the number in parentheses) were estimated to be above the estimated standard. The reason for this difference in percentages is that a sizable portion of the examinees who were classified as below the standard by the panelists were above the cut scores that were estimated by the different methods. Figure 8.7 shows this graphically.

The information in Figure 8.7 shows that the distributions of the examinees classified above and below the cut score by the panelists have substantial overlap. Further, of the examinees classified by panelists as below the standard, 68% or 126 are above the estimated cut score of 1.28 estimated from the point of intersection of the two distributions. The percent of the distribution above the cut score for those classified as above by panelists is 95% or 535 examinees. Together,

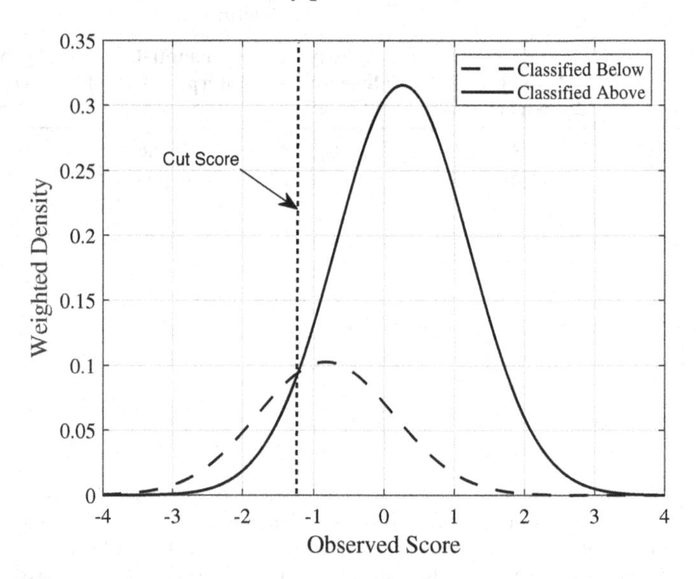

FIGURE 8.7
Observed score distributions based on panelists classifications.

there are 661 examinees above the cut score on the observed score scale, or 88%. That is the case even though the panelists classified only 565 candidates as above the cut score. A comparison of Figure 8.7 with Figure 8.6 shows the impact of panelist classification reliability on the overlap of the distributions. It is interesting to note that compared to the results shown in Table 8.2, the percent above the standard on the observed score scale is in the same range. Because this is based on a simulation, it is also possible to check the classifications based on the estimated standard and the observed score against the true classifications based on the true scores on the construct defined by the agency. The results of that comparison show that using the estimated standard and observed scores resulted in about 6% misclassifications with about 5% in the positive direction. This might be considered an acceptable error rate given the amount of error in the standard setting process.

Table 8.10 contains the summary statistics for the implementation of the Contrasting Groups kernel with 15 panelists all classifying the same set of work samples. The standards based on the observed data from that procedure

TABLE 8.10

Descriptive Statistics for the Four Scales Used When Generating the Simulated Data for Standard Setting with All Panelists Rating the Same 100 Work Samples

	Scale Definition			
Statistic	**Agency True Construct Scale**	**Agency Observed Scale**	**Panelists' Mean Interpretation of Agency Scale**	**Panelists' Mean Observed Scale**
Mean	−.0847	−.0799	−.0912	−.1198
Standard deviation	1.0114	1.0560	.9275 (1.0222)#	.9271 (1.3794)#
Proportion above standard	.84	.80	.86	.81 (.86 to .88)*
		Correlations		
Agency true		.97	.99	.98
Agency observed	.9450		.95	.95 (.6593)+
Panelists' interpretation	.9928 (.9080)+	.9360		.99
Panelists' observed	.9554 (.6749)+	.8987 (.6359)+	.9649	

Notes: The only data that are observable during a standard setting study are the shaded cells. The values in parentheses (*) give the range of proportions above the estimated standard using data from the Agency Observed Scale. The values in parenthesis (#) are the pooled standard deviations instead of the standard deviation of the mean estimates. Correlations below the diagonal are Pearson Product Moment correlations between locations of examinees on the score scales. Those in parentheses (+) are the average correlation with individual panelists scales. Correlations above the diagonal are tetrachoric correlations between classifications relative to the cut scores.

and the observed scores from the simulated examinees yield a similar percent passing to the procedure having each panelist classify 50 candidates and then aggregate the results, but the tetrachoric correlation of the final classification of candidates with the true classification from the agency's true construct is much higher when a common set of work samples are classified (.98) than when each panelist classified their own smaller set of candidates (.60). This large difference shows the advantage gained by averaging results over panelists (in this case the modal classification was used) rather than having panelists classify independent groups and using a larger number of candidates (750 versus 100).

It is difficult to generalize the results from these simulations to all possible implementations of the Contrasting Groups kernel. However, the simulation does show the components that are underlying this method and some of the standard setting procedure design features that can influence the accuracy of the results. Those who design and implement standard setting procedures should conduct their own research studies to determine how to design the procedures to give accurate estimates of the cut scores intended by the agency.

8.3.2 Estimation Based on Test Item Judgment Data

The previous section of this chapter discusses the estimation of a standard when the information from panelists (translators) is classifications of persons or work samples into groups. The examples are useful because they indicate how test reliability and judgment error influence the accuracy of estimates of the intended standard. The discussion in this section extends the previous examples to the case when panelists are asked to make judgments about what a person at the standard can do on test items rather than about whether the persons or work samples are representative of the range above or below the standard. The goal is the same. It is to estimate the location of the intended standard on the construct defined by $\mathbf{D}_\mathbf{P}$. The difference is that the panelists now use their conception of the standard to guide judgments about how a person at the standard will perform on specific test tasks.

Suppose that all the information provided for the previous example still holds. An agency has defined a policy $\mathbf{D}_\mathbf{P}$ and from that policy developed $\mathbf{D}_\mathbf{C}$ and $\mathbf{minD}_\mathbf{C}$. Further, there is a construct implied by $\mathbf{D}_\mathbf{P}$ that is indicated by ζ and there is an intended standard ζ_{Ag} on the implied construct of –1. The goal of the standard setting process is to get a good estimate of ζ_{Ag} from the information obtained from panelists in a standard setting study. In this case, the kernel for the standard setting process is a method for collecting information about how individuals whose location on the implied construct is at –1 are expected to perform on the test items. The goal is to get a good estimate of the intended standard from the information provided by panelists. That information is the content or **R**.

One additional assumption is needed to consider the possible approaches to estimating the intended standard. That assumption is that the relationship between the location of candidates on the hypothetical construct and the probability that they answer each test item correctly can be expressed using a mathematical function of item characteristics and a candidate's location. In this case, it is assumed that the relationship between item and person characteristics and the probability of a correct response to the test items is accurately represented by the three-parameter logistic model (Birnbaum, 1968). This model is expressed mathematically as follows:

$$P\left(u_{ij}=1 \middle| \theta_j\right)=c_i+\left(1-c_i\right)\frac{e^{Da_i\left(\theta_j-b_i\right)}}{1+e^{Da_i\left(\theta_j-b_i\right)}}, \tag{8.1}$$

where

- a_i is a parameter describing the discriminating power of the item (also called the slope parameter);
- b_i is a parameter describing the difficulty of the item (also called the location parameter);
- c_i is a parameter related to the likelihood of guessing the correct response to the item (also called the lower asymptote parameter);
- D is a constant approximately equal to 1.7 that scales the function to be similar to the normal ogive function;
- θ_j is a parameter indicating the level of proficiency of the examinee;
- u_{ij} is the score on the test item with 1 indicating a correct response and 0 an incorrect response.

The example provided here is based on an eighth-grade mathematics test from a state summative assessment examination program. However, all the item parameters were simulated by randomly sampling from distributions that had the same means and standard deviations as the parameters from an actual test. The *a*-parameters were sampled from a log-normal distribution, the *b*-parameters were sampled from a normal distribution, and the *c*-parameters were sampled from a beta distribution. Table 8.11 contains the parameters and summary statistics, and Figure 8.8 gives a graphic representation of the assumed mathematical function for Item 1 on the test. The graph of the mathematical function is typically labeled as an item characteristic curve.

Figure 8.8 shows the properties of Item 1 when the mathematical model given in Equation 8.1 is an accurate representation of the relationship between the probability of a correct response to a test item and the proficiency level of the examinees. First, the slope of the curve is steepest above the point on the construct equal to the *b*-parameter of –.76. At that point on the scale, the probability is equal to $(1 + c)/2$. In this case, the probability is .6. Further, the curve has a lower asymptote at the value of c and an upper asymptote of 1.0. The curve is also monotonically increasing. Each item in the test has a curve of this type with features defined by the parameters listed in Table 8.11. Note that the probability of correct response for Item 1 at the intended standard of –1 is .5288. This is an important value for several standard setting kernels

TABLE 8.11

True Simulated Item Parameters for a 60-Item, Eighth-Grade Mathematics Test

Item Number	a_i	b_i	c_i	Item Number	a_i	b_i	c_i
1	0.91	−0.76	0.20	31	0.98	−0.80	0.21
2	1.20	−0.09	0.14	32	0.63	−0.05	0.16
3	0.50	0.22	0.21	33	0.65	0.33	0.21
4	0.98	0.55	0.14	34	0.68	1.44	0.23
5	0.87	0.82	0.10	35	0.43	−0.51	0.17
6	0.61	−0.06	0.15	36	1.10	0.00	0.16
7	0.74	−1.00	0.20	37	0.87	−0.16	0.19
8	0.87	−0.55	0.19	38	0.69	−1.27	0.29
9	1.74	−0.74	0.31	39	1.09	−0.37	0.24
10	1.47	1.30	0.17	40	0.56	−1.18	0.20
11	0.61	−0.48	0.21	41	0.79	0.40	0.16
12	1.55	0.34	0.18	42	0.77	−0.64	0.20
13	0.95	−0.22	0.18	43	0.87	−0.05	0.20
14	0.80	0.43	0.19	44	0.87	−0.43	0.16
15	0.95	−0.57	0.23	45	0.68	0.07	0.23
16	0.78	−0.95	0.18	46	0.81	−0.47	0.29
17	0.79	−0.96	0.13	47	0.78	0.19	0.17
18	1.12	0.18	0.18	48	0.93	0.34	0.14
19	1.10	−0.21	0.29	49	1.03	0.92	0.14
20	1.10	−0.23	0.16	50	1.02	−0.22	0.23
21	0.94	0.74	0.13	51	0.68	−1.39	0.31
22	0.63	0.07	0.30	52	0.83	−0.61	0.17
23	0.95	0.01	0.18	53	0.62	0.70	0.17
24	1.15	0.84	0.16	54	0.64	−0.75	0.21
25	0.90	−0.59	0.20	55	0.81	0.47	0.19
26	1.01	0.31	0.09	56	1.13	−0.03	0.12
27	0.95	0.39	0.18	57	0.69	0.75	0.17
28	0.76	−0.25	0.21	58	0.88	−1.28	0.14
29	0.86	0.02	0.26	59	0.77	−0.23	0.07
30	0.69	−0.81	0.29	60	1.03	−0.83	0.24
Mean					0.88	−0.13	0.19
Standard deviation					0.24	0.64	0.05

that are based on panelists' use of item judgments to arrive at a translation of $\mathbf{D_P}$ to a point on the construct implied by the definition. Three standard setting kernels will be considered here: the Angoff probability-estimation kernel, the Angoff item-score kernel, and the Bookmark kernel. The accuracy of estimation of the intended point on the reporting score scale will be evaluated for each of these kernels and several models for item parameter estimation error and panelists' judgment error.

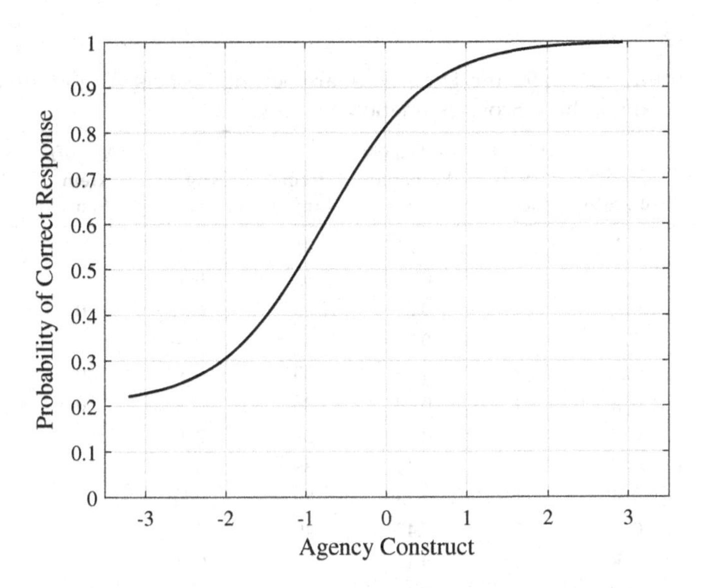

FIGURE 8.8
Item characteristic curve for Item 1: $a = .91$, $b = -.76$, $c = .20$.

Before investigating the influence of error on the different standard setting kernels, the estimation of the intended standard is shown under the ideal conditions that there is no error in item parameters and the panelists make the judgments required by each kernel without error. Table 8.12 contains the information in **R** for each of the standard setting kernels for this ideal condition. This information includes the probability of correct response to each item for an examinee with proficiency equal to the intended standard of –1 on the construct defined by $\mathbf{D}_P$, the most likely item score for each item for an examinee at the intended standard with a 1 assigned if the probability of correct response is greater than .5 and 0 otherwise, and the probability of correct response for an examinee at the intended standard with the influence of the *c*-parameter removed for use with the bookmark procedure. The bottom row of the table gives the sum over items of the values in the table. These are estimates of the total score on the test that an examinee at the intended standard would be expected to obtain under the conditions for each standard setting kernel. The sum for the Bookmark kernel is the score expected for an examinee at –1 on the construct if there was no possibility of responding to the test items correctly using anything other than the skills and knowledge implied by $\mathbf{D}_C$ (e.g., no guessing).

The item order in Table 8.12 follows the approach for creating the ordered item booklet (OIB) for the Bookmark kernel. Items were ordered from easiest to the most difficult according to the probability adjusted to remove the influence of the *c*-parameter. If the standard setting procedures followed by Mitzel et al. (2001) are followed to estimate the location of the standard, the bookmark would be placed on Item 2 in the OIB, the last item with a

TABLE 8.12

Error-Free Information for the Three Standard Setting Kernels: Angoff Probability Estimation, Angoff Item-Score Estimation, Bookmark

Item Number*	Angoff Probability	Angoff Item Score	Bookmark Adjusted for c_i	Item Number	Angoff Probability	Angoff Item Score	Bookmark Adjusted for c_i
1	0.75	1	0.71	31	0.37	0	0.22
2	0.78	1	0.69	32	0.36	0	0.22
3	0.76	1	0.66	33	0.39	0	0.20
4	0.67	1	0.59	34	0.36	0	0.20
5	0.65	1	0.56	35	0.32	0	0.19
6	0.56	1	0.49	36	0.36	0	0.19
7	0.58	1	0.48	37	0.42	0	0.19
8	0.61	1	0.44	38	0.40	0	0.18
9	0.55	1	0.43	39	0.31	0	0.17
10	0.57	1	0.43	40	0.32	0	0.16
11	0.54	1	0.42	41	0.29	0	0.14
12	0.51	1	0.41	42	0.24	0	0.14
13	0.53	1	0.41	43	0.26	0	0.13
14	0.51	1	0.38	44	0.27	0	0.13
15	0.50	1	0.37	45	0.27	0	0.13
16	0.48	0	0.37	46	0.29	0	0.13
17	0.48	0	0.35	47	0.29	0	0.12
18	0.47	0	0.34	48	0.26	0	0.11
19	0.49	0	0.33	49	0.23	0	0.11
20	0.52	1	0.33	50	0.26	0	0.10
21	0.53	1	0.32	51	0.25	0	0.10
22	0.42	0	0.30	52	0.18	0	0.09
23	0.43	0	0.28	53	0.20	0	0.07
24	0.38	0	0.27	54	0.16	0	0.06
25	0.32	0	0.27	55	0.18	0	0.06
26	0.38	0	0.26	56	0.28	0	0.06
27	0.42	0	0.26	57	0.16	0	0.03
28	0.47	0	0.24	58	0.20	0	0.03
29	0.42	0	0.24	59	0.18	0	0.03
30	0.41	0	0.23	60	0.17	0	0.00
Sum					23.89	17	15.54

probability greater than .67. This bookmark placement is very early in the OIB because the test used as the model for this simulation was designed to have four cut scores representing increasing levels of performance. The intended standard used for the simulation is the lowest of the four standards.

Figure 8.9 shows a magnified section of the item characteristic curves for the second and third items in the OIB along with the .67-line that was used for mapping the curves to the scale of the construct implied by $\mathbf{D}_P$. The two

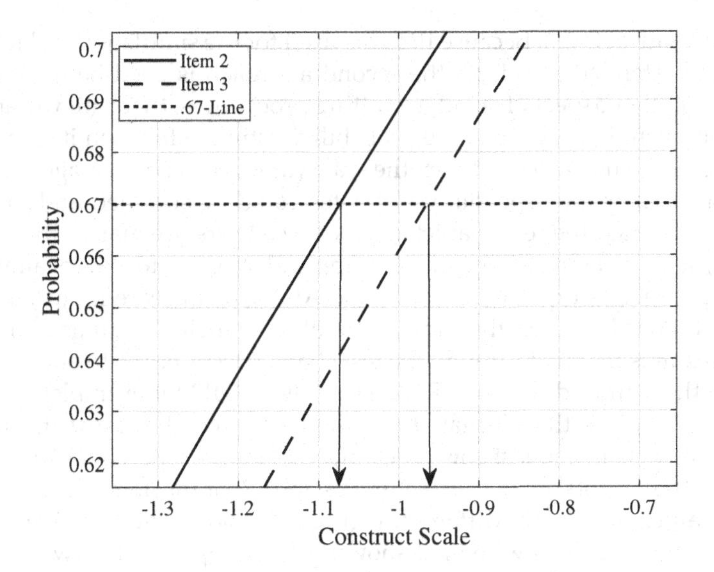

FIGURE 8.9
Mapping of the bookmarked and the following item onto the scale for the construct assessed by the test.

arrows show where the items were mapped to the scale to obtain an estimate of the intended standard. The bookmark was placed on Item 2 as the last item in the OIB with a probability of correct response for a person described by $\mathbf{minD_C}$ that is greater than .67. That item maps to –1.07 on the construct for the test. Item 3 maps to –.96 on the scale. These values are very close together on the scale because in this instance there is only a small difference in the difficulties of the two items. Neither value is exactly equal to the intended standard of –1 because there is a gap between the two items. However, the midpoint between the two values, –1.02, is an estimate of the standard that is very close to the intended standard. Even under these ideal conditions of no error in the item parameters and panelist's judgments, the bookmark procedure will not exactly recover the intended standard unless a test item has an item with exactly a .67 probability for persons at the standard. However, the Bookmark kernel will give a value close to the intended standard if the midpoint of the mapped locations of the items on either side of the standard is used as the estimate. An important question is how robust the estimation procedure is to errors in the item parameters and in the judgments made by panelists.

The estimates of the standard using the two Angoff kernels, probability estimation and item-score estimation, are usually made using one of two methods. One is to simply sum the estimates for each item given by the panelists to get an estimate of the standard on the sum-score scale for the test. The second approach is to map the sum of the panelists' estimates through the test characteristic curve for the test to obtain an estimate of the standard on the item response theory (IRT)–based proficiency scale (see, for example, Clauser,

Kane, & Clauser 2020). Because IRT was used for the simulation and for specifying the intended standard, the second approach is used here. Of course, like every other aspect of standard setting procedures, there are variations in that approach. For example, the probability estimate for each item could be mapped to the IRT scale and then the scale values could be averaged to obtain the standard, or the probability estimates could be summed and the result mapped through the test characteristic curve. There are many other options as well, so no attempt will be made to apply all of them for this example. The same approach is typically used for the Angoff item-score kernel. The estimated standard is either the simple sum of the panelist's ratings of the items or the result is mapped through the test characteristic curve to obtain an estimate of the standard on the IRT scale (see Hsieh 2013 for example).

Table 8.12 gives the estimated summed score on the test using the two Angoff kernels in the last row. The values of sum of the item estimates are 23.89 for the Angoff probability estimates and 17 for the item-score estimates. These values are clearly different. Figure 8.10 shows the test characteristic curve for the test along with lines showing the mapping of the two estimated scores on the test. The figure also shows the curve for the sum of the most likely item scores. This curve was computed by rounding the conditional probability of correct responses to the items to 0 or 1 and then summing them. This curve is consistent with the Angoff Yes/No kernel.

If the test characteristic curve (TCC) as is typically defined by IRT texts is used for the mapping, the value of 23.89 from the Angoff probability estimates exactly recovers the intended standard of –1 for the ideal case described here. This result is expected because the sum or the probability

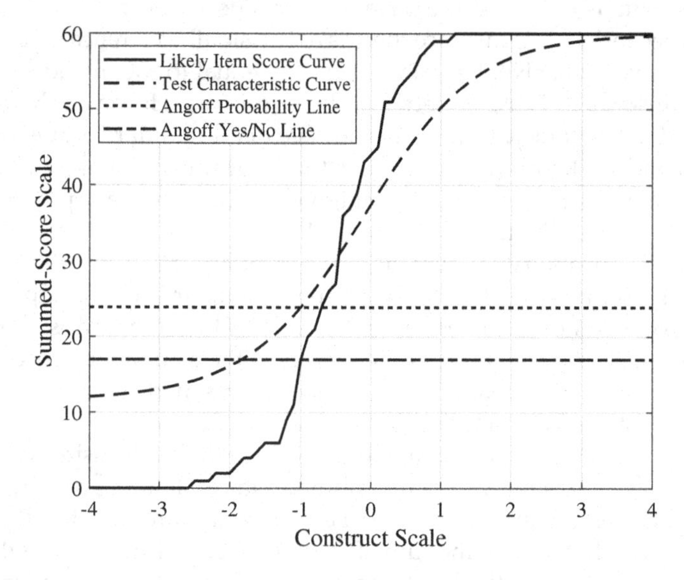

FIGURE 8.10
Item response theory curves for mapping the panelist's ratings to the IRT scale.

estimates obtained without error are an estimate of the expected score on the test and the TCC is defined as the relationship between the IRT scale value and the expected score on the test. The value obtained from mapping the 17 obtained from the Angoff item-score estimates through the test characteristic curve is –1.84, much lower than the intended standard of –1.

Some reflection on the difference obtained from these two methods indicates that the difference is because the two standard kernels are asking for different things from the panelists. The Angoff probability-estimation kernel is asking the panelists to give the expected score on each test item conditional on $\mathbf{minD}_C$. The sum of the expected scores on the items gives the expected sum-score on the test because the expected value of a sum is the sum of the expected values.

$$E\left(x_j \middle| \mathbf{minD}_C\right) = E\left(\sum_{i=1}^{n} u_{ij} \middle| \mathbf{minD}_C\right) = \sum_{i=1}^{n} E\left(u_{ij} \middle| \mathbf{minD}_C\right) \tag{8.2}$$

$$= \sum_{i=1}^{n} \left[\left(u_{ij} = 1\right) * P\left(u_{ij} = 1 \middle| \mathbf{minD}_C\right) + \left(u_{ij} = 0\right) * P\left(u_{ij} = 0 \middle| \mathbf{minD}_C\right)\right]$$

$$= \sum_{i=1}^{n} P\left(u_{ij} = 1 \middle| \mathbf{minD}_C\right)$$

where

x_j is the summed score on the test for examinee *j*;
n is the number of items on the test;
u_{ij} is the score on item *i* for examinee *j*;
$E(\,)$ is the expected value operator;
$P(\,)$ is the probability computed from the IRT model.

The sum of the Angoff item-score estimates does not have the same relationship with the expected score on the test, so mapping through the TCC is not appropriate. Rather than asking panelists to provide the expected score on the item, the Angoff item-score kernel asks the panelists to provide the most likely score. The usual statistical interpretation of "most likely" is the score on the item that has the higher (highest) probability of occurrence. In the simple case of items scored 0 and 1, the panelist is asked which of the scores has a probability greater than .5 for a person who has the characteristics given in $\mathbf{minD}_C$. For the case of items scored 0 and 1, the cut score for the Angoff item-score kernel can be computed by rounding the probability of correct response to 0 or 1:

$$y_j = \sum_{i=1}^{n} \text{round}\left(P\left(u_{ij} = 1 \middle| \mathbf{minD}_C\right)\right), \tag{8.3}$$

where y_j is the score for Person j who matches the requirements of $\mathbf{minD_C}$, and the other terms have their previous definitions. The value of y_j in Equation 8.3 can be computed for possible values of the cut score on the intended construct yielding a curve that is roughly equivalent to the test characteristic curve. This curve is labeled the "likely item-score curve" (LISC) showing the relationship between the intended standards on the construct defined by $\mathbf{D_P}$ and the score obtained from the standard setting kernel, in this case the Angoff item-score kernel. The LISC is the "Likely Item-Score Curve" in Figure 8.10. When this curve is used to map the value of the standard on the number-correct score scale (i.e., 17) to the scale of the construct, it again recovers the intended standard of –1. This suggests that if the Angoff item-score kernel is used for setting standards on an IRT-based score scale, the mapping of the estimated summed score to the IRT scale should be through the LISC for that standard setting kernel rather than the TCC.

An important result can be observed from the comparison of the TCC and LISC in this case. For every possible standard on the construct defined by $\mathbf{D_P}$ except where the two curves cross, the LISC shows that the summed score value obtained from the Angoff item-score kernel is more extreme than that for the Angoff probability kernel. The difference in the curves can have dramatic results if the estimated standard on the summed score scale is used for decision-making. In this example, the Angoff probability kernel yields a summed score with 85% of the score distribution above it, consistent with the value of –1 on the construct scale. The Angoff item-score kernel yields a summed score with 96% above it, a value that is much greater than intended.

A specialized curve can also be developed for the Bookmark kernel (bookmark characteristic curve, BCC). Under the ideal conditions of no error in calibration of the items and no error in placement of the bookmark, the BCC gives the number of test items before the bookmark, including the bookmarked item conditional on possible values of the intended standard. Figure 8.11 shows the Bookmark kernel BCC along with specialized curves for the other two standard setting kernels.

It is important to note that the three curves in Figure 8.11 all intersect the lines derived from the information obtained from panelists at –1 showing that all three methods could recover the intended standard if the appropriate curve were used for mapping the information to the scale. However, Figure 8.11 also shows that the curves are quite different so that if the incorrect curve were used, the estimated standard would be far from the intended standard. Also, the curves differ in their slopes indicating that for the same amount of change on the summed score scale, there are different amounts of change on the construct implied by $\mathbf{D_P}$. This has implications for the sensitivity of the standard setting kernels to feedback provided to panelists between rounds of the standard setting process.

The differences in the TCC, the LISC, and BCC suggest that different standard setting kernels require different estimation procedures. The three curves shown here can be generically labeled as kernel characteristic curves

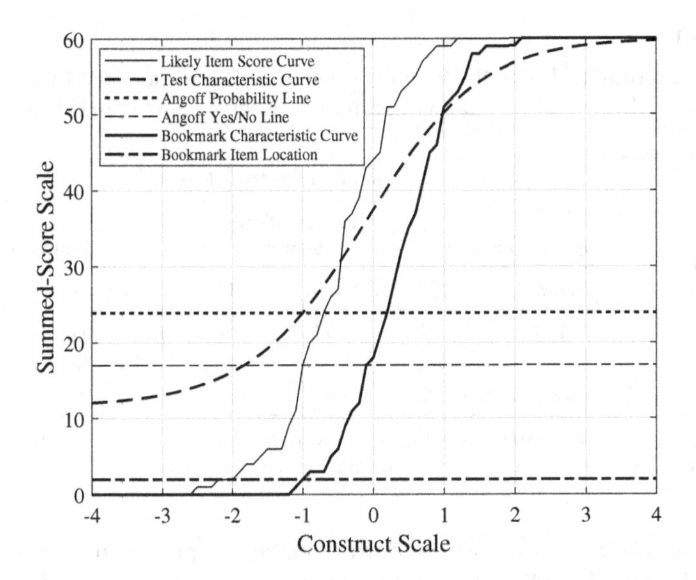

FIGURE 8.11
Kernel characteristic curves for mapping the panelist's ratings to the IRT scale.

(KCCs) and each standard setting kernel can have its own version of a KCC. It would be very informative if reports of standard setting processes would include the KCC for the standard setting kernel that was implemented for the standard setting process.

8.3.2.1 Effect of Item Parameter Estimation Error on Estimates of Standards

The information in the previous section assumes accurate judgments for the tasks required of the standard setting kernel and accurate item parameters for the items on the test. The results show that if the proper KCC is used to map the results in **R** to the scale for the intended construct for the standard, the two Angoff kernels exactly recover the intended standard and the Bookmark kernel gives a very close estimate, differing from an exact estimate because of a small gap between items mapped to either side of the intended standard.

These results support the use of those standard setting kernels, but they are far from realistic. Panelist judgments are not precise and item parameters contain estimation error. It is important to understand how the various kinds of error influence the estimation of the standard. The first type of error to be investigated is the amount of error in the estimation of item parameters. From a review of the literature, the amount of estimation error for large sample calibration of items modeled with the three-parameter logistic model is approximately a standard error of .1 for the *a*-parameters, .09 for the *b*-parameters, and .03 for the *c*-parameters. These values for the standard errors of the parameters were used to add an error term to the parameters by randomly sampling a value from distributions

TABLE 8.13

Standards Estimated Using the Item Parameters with Error for the Three Standard Setting Kernels: Angoff Probability Estimation, Angoff Item-Score Estimation, Bookmark

	Standard Setting Kernel					
	Angoff Probability		Angoff Item Score		Bookmark	
Intended Standard	Mean	SD	Mean	SD	Mean	SD
–1	–1.01	.02	–.97	.05	–1.12 (–1.03)	.13 (.10)
0	.00	.02	–.02	.05	–.13 (–.07)	.13(.11)
1	1.01	.02	1.03	.08	.94 (1.03)	.14 (.14)

Note: The values in the parentheses for the Bookmark kernel are the mean estimates for the midpoint between the locations of first and second bookmarks.

with means equal to 0.0 and standard deviations equal to those approximate standard errors. Then the estimation of the standard was conducted using accurate estimates of the data in **R** but using the parameters with error to get the estimated standard. This process was replicated 1,000 times to get a sampling distribution of the standard. Table 8.13 gives the mean and standard deviations of the estimates for each kernel assuming an intended standard of –1 on the intended construct.

When these simulations were performed, when item parameter estimation error was included in the data for the Bookmark kernel, sometimes the very first item in the OIB did not reach the criterion of a probability of .67 for examinees described by $\mathbf{minD_C}$. The low probability was a result of using parameters with error. The observed OIB had a slightly differing ordering of items than that based on parameters without error. About 10% of the time for the test modeled here, the item ordered in the first position was a more difficult item than would have been the case based on true item parameters. In theory, that suggests a panelist who accurately judged the item would indicate that the bookmark would be placed before the first item. In this case, the estimated standard was computed as the mapping location for the first item minus .3 on the θ-scale. This is a consequence of the intended standard used for this analysis being the lowest of several standards and the test not being well matched to that level of performance. Because of concerns that this might distort the results, the simulations were also run for intended standards of 0 and 1 on the construct defined by $\mathbf{D_P}$. Table 8.13 includes those results along with the results for –1 on the scale. The results are based on 1,000 replications with different sampling of item parameter estimation error for each replication.

The results show that the effect of item parameter estimation error was small for the two Angoff kernels. The mean estimates were close to the intended standards and the standard deviations were small. It is important to note that the estimates of standards for the Angoff item-score kernel were based on mapping the summed score through the NISC instead of the TCC. If the TCC had been used, the estimates for the Angoff item-score kernel would have been in the region of –1.84 for the intended standard of –1. It is important that estimation of the standard for the Angoff item-score kernel uses the NISC.

The Bookmark kernel gave consistently lower estimates of the intended standard and had larger standard deviations of the distributions of the estimates than the other methods. These findings are a result of ordering of test items in the OIB based on the item parameters with error and computing the location of the item with the bookmark based on the parameters with error. The result is that the simulated panelist often correctly identifies that an item has a probability of correct response that is below .67 earlier in the OIB than would be the case if the true item parameters were used for the ordering, and then the mapped location of that item is based on the item parameters with error for that item.

This underestimation of the intended standard using the Bookmark kernel might be reduced if panelists were instructed to find the first item that has a probability less than .67 for an examinee described by $\mathbf{minD}_C$ and then look to see if there is an item further in the OIB that they judge to have a probability above .67. The estimate of the standard could be halfway between the mapped locations for these two items. This approach was applied to the simulated data and Table 8.13 presents the results in parentheses in the Bookmark columns. The results show that this process reduces the amount of underestimation, but the standard deviations of the distributions of the estimated standards (e.g., standard errors) are still greater than those for the Angoff kernels.

The positive outcome of this analysis of these standard setting kernels is that there does not seem to be much impact of the amount of item parameter estimation error that was included in the simulation. The amount of error modeled was consistent with the estimation error from calibrations from large-scale assessment programs. These programs have large and representative samples of examinees. These results may not generalize to other cases with small, nonrepresentative samples.

8.3.2.2 Effect of Panelists' Errors in Estimation of Conditional Probabilities

The amount of error in panelists' contributions to $\mathbf{R}$ is of greater concern than the error in item parameter estimates. The calibration of items is typically done on a large sample yielding small errors. As the previous section shows, those errors do not seem to have a large impact on the estimates of the intended standard. On the other hand, there is a body of research going as far back first half of the 19th century (Thorndike, Bregman & Cobb, 1924; Tinkelman, 1947) that show that judgments of item difficulty are reasonably correlated with empirical estimates indicating that panelists can likely order items according to their difficulty with some level of accuracy, but that estimates of the actual value of a difficulty measure is not very accurate. Because the Angoff and Bookmark kernels are dependent on the accuracy of panelists' estimates of the probability of a correct response to items, it is important to check the sensitivity of the estimation of the intended standard when there is a reasonable amount of error in panelists' estimates of the probability of correct response for persons who match $\mathbf{minD}_C$.

The simulations used in the previous sections of this chapter were extended to include a reasonable amount of error in the panelists' estimates of the conditional probabilities. The error was modeled to have a beta distribution with a mean equal to the true conditional probability obtained from the IRT model used to generate the data and standard deviation of .1 when the conditional probability is .5. The standard deviation of the distribution of errors was less as the conditional probability approaches 0 or 1 because of the restricted range of the probability scale.

The results of this simulation yield some interesting and quite dramatic results (see Table 8.14). First, the standard deviations (estimates of standard errors of the estimated standard) are roughly twice as large for each of the kernels when compared to the case when there are item parameter errors, but no panelist judgment errors. This is expected because an additional source of error has been added for this analysis. The surprising result is the underestimation of the intended standard by the Bookmark kernel. The underestimation increases as the level of the intended standard increases with the largest amount of underestimation when the intended standard was 1. The reason for that underestimation is that as panelists work through the OIB from the easiest item to the most difficult item, the likelihood that an error in judgment will result in placing the bookmark earlier in the OIB increases. If the panelists were placing the bookmark for the intended standard of 1 accurately, they would place it on page 51 resulting in an estimated estimate of .98. However, the errors in the judgment of conditional probabilities results in a median page number of 38, much earlier in the OIB than the bookmark

TABLE 8.14

Standards Estimated Using Item Parameters with Error and Panelist Error for the Three Standard Setting Kernels: Angoff Probability Estimation, Angoff Item-Score Estimation, Bookmark

	Standard Setting Kernel					
	Angoff Probability		Angoff Item Score		Bookmark	
Intended Standard	Mean	SD	Mean	SD	Mean	SD
–1	–1.01	.07	–.97	.09	–1.16 (–1.01)	.23 (.21)
0	.00	.05	–.04	.11	–.38 (–.32)	.27(.22)
1	1.01	.06	.90	.19	.52 (.56)	.29 (.25)

Note: The values in the parentheses for the Bookmark kernel are the mean estimates for the midpoint between the locations of first and second bookmarks.

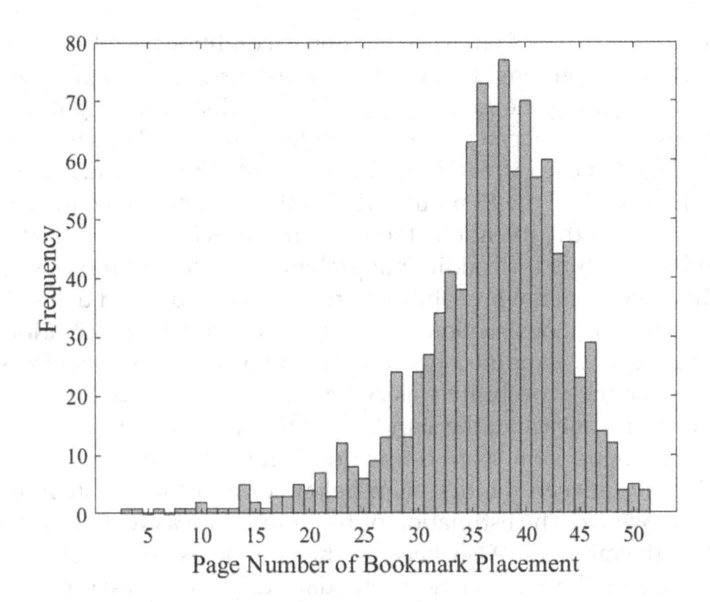

FIGURE 8.12
Distribution of bookmark placements for intended standard of 1 with panelist estimation error and item parameter error.

consistent with the intended standard. Figure 8.12 shows the distribution of bookmark placements for an intended standard of 1 when there is error in both the item parameters and the panelist judgments.

While the Bookmark kernel yields estimates of the intended standard that are below the actual value and they also have large standard errors, the Angoff probability kernel resulted in very good estimates with relatively small standard errors. While these results are very encouraging, they should be interpreted with caution because they assume that the distribution of panelists' estimates of conditional probability have a mean equal to the probability that is consistent with the true standard. If panelists have some systematic error in their estimates, such as regressing the estimates toward the center of the probability scale, the results would not be as positive.

The Angoff item-score estimation kernel gave reasonable results when the sum of the correct/incorrect scores were mapped to the IRT scale using the NISC. That kernel has larger standard errors than the Angoff probability kernel and some slight regression of estimates, but nothing of the magnitude observed for the Bookmark kernel.

8.3.2.3 Including Panelists' Interpretation of D_c and minD_C in the Estimation of the Intended Standard

Section 8.3.1 showed how differences in panelists' interpretation of the definitions influence the estimation of the intended standard. So that the results from these kernels based on the estimation of conditional probabilities of correct response can be compared to those earlier results, the simulations were conducted assuming the same group of 15 panelists with slightly different interpretations of the location of the intended standard. The same variations in interpretations used in Section 8.3.1 were used in this simulation.

In this case, each of the 15 simulated panelists had their own interpretation of the location of the intended standard on the IRT scale and that location was used to compute the conditional probability of correct response for each item. Then a conditional probability with error was obtained using the same process as the previous section—sampling from a beta distribution with a mean equal to the computed probability and a standard deviation equivalent to .1 when the probability was .5. The probabilities with error were then used to implement each of the standard setting kernels—Angoff probability estimation, Angoff item-score estimation, bookmark, and bookmark using the midpoint between the first item below .67 and the last item identified that was above .67. The estimation of the standard was replicated ten times for each of the methods. Also, because the intended standard of 1 gave such a different result than the intended standard of –1 in the last section, both of those values were used as the intended standard in this simulation.

Table 8.15 presents the results of the simulation. These results show that for the conditions used for the simulation, the Angoff probability kernel resulted in accurate and stable estimates of intended standards at both levels. The Angoff item-score kernel using the NISC for estimating the standards gave accurate results for the intended standard of –1 but underestimated the standard at 1 and had a larger standard error of the estimate at that level. The Bookmark kernel using both the initial bookmark placement and the mean of two bookmark placements greatly underestimated the standard at 1 but gave more accurate results for the lower standard. The mean of the two bookmarks was more accurate than the location of the initial bookmark.

TABLE 8.15

Standards Estimated Using Item Parameters with Error, Panelists Interpretation of the Intended Standard, and Panelist Judgment Error for the Three Standard Setting Kernels: Angoff Probability Estimation, Angoff Item-Score Estimation, Bookmark

	Standard Setting Kernel					
	Angoff Probability		Angoff Item Score		Bookmark	
Intended Standard	Mean	SD	Mean	SD	Mean	SD
–1	–1.03	.01	–.98	.02	–1.17 (–1.04)	.05 (.03)
1	1.02	.01	.92	.07	.55 (.59)	.07 (.07)

Notes: The values in the parentheses for the Bookmark kernel are the mean estimates for the midpoint between the locations of first and second bookmarks. Results are based on ten replications of the process.

These results assume that both the estimates of the item parameters and the panelists' estimates of probabilities are unbiased—the mean of the estimates is equal to the true value. Under that assumption, the Angoff probability-estimation kernel is preferred because it closely recovers the intended standard with small standard error. This result should not be interpreted as being too definitive because there is some evidence that panelists regress their estimates of probability toward the center of the probability scale (Wyse, 2018). It is interesting that a regression effect has a strong effect on the Angoff probability-estimation and Bookmark kernels, but little effect on the Angoff item-score kernel if the probabilities are regressed toward .5. In the Angoff item-score kernel simulation, conditional probability estimates greater than .5 were assigned a score of 1 and those below .5 were assigned a score of 0. Regressing the probabilities toward .5 does not change these score assignments because no probability estimates cross over the .5 point on the probability scale. If the probabilities were regressed toward a different probability, such as .7, then there would be a change in the estimated standard for all the kernels that were investigated here.

8.3.3 Estimation Using Compromise Methods

There are several methods in the standard setting literature that are sometimes called "compromise methods" (Cizek & Bunch, 2007). These methods were given that label because they use a combination of what are considered criterion-referenced and norm-referenced methods. They are methods that ask panelists to use two different standard setting kernels and then use quasi-statistical procedures to combine the results of the two kernels to obtain an estimate of the standards. The two examples of these kinds of methods are the Hofstee (1983) method and the Beuk (1984) method.

The first standard setting kernel used by the Hofstee (1983) method is direct estimation of the standard on the percent correct reporting score scale. To account for error in panelists' judgments, Hofstee asked panelists to respond to two questions: "What is the highest percent correct cut score that would be acceptable, even if every examinee attains that score?" "What is the lowest percent correct cut score that would be acceptable even if no examinee attains that score?" (Cizek & Bunch, 2007; p. 210). The second standard setting kernel is the direct estimation of the percent below the cut score. From such values and the empirical distribution of scores, the percent below the cut score can be mapped through the cumulative distribution function of the scores to identify the point on the reporting score scale that corresponds to the standard. They ask two questions of the panelists related to this second standard setting kernel: "What is the maximum acceptable failure rate?" "What is the minimum acceptable failure rate?"

It is not clear how panelists would arrive at the answers to the questions given their understanding of the policy and content definitions. The first kernel requires that panelists have a good understanding of the intended construct and the characteristics of the test that is used to operationally define the approximation to the intended construct—the reporting score scale. It is possible that panelists with extensive experience using the test results, such as admissions officers at universities, have a good understanding of the level of performance related to the intended construct for various points on the reporting score scale. Those persons might be able to provide the information required by the first kernel with an acceptable level of accuracy. But the understandings of these panelists might be connected to scale scores from the test rather than percent correct scores. It is seldom that a large-scale or high-stakes test reports results on a percent correct scale. Using a scale score facilitates equating multiple test forms and discourages interpretations that do not adequately adjust for the difficulty of the test form such as using 70% correct is passing. There does not seem to be anything in the Hofstee (1983) method that precludes using scale scores as the basis for the first kernel of the process, but a search of the literature on the use of the Hofstee method did not result in any examples that used anything other than the percent correct score with the method.

The second kernel in the Hofstee (1983) method is a direct estimate of the proportion of examinees who will score above the intended cut score. Panelists need to have a good understanding of $\mathbf{D_P}$ and the distribution of performance of the sample of examinees who typically take the test to provide a reasonable estimate of the required proportions. Panelists with extensive experience with the test may have a good understanding of the characteristics of the observed score distributions and the typical examinee sample. The panelists also need to have a policy understanding of the acceptable failure rate. There is substantial complexity when estimating acceptable failure rates because it involves consideration of the location of the intended cut score and many other educational and social factors. For example, there may be limited capacity for training programs in a specialty area and so an acceptable failure rate might be very high as a practical necessity rather than being related to a person's location on the intended construct. Such situations have led to lottery systems for selecting candidates rather than building information about instructional management systems into estimating the cut score.

Hofstee (1983) uses the answers to the questions about performance on the test and proportion failing to define the set of combinations of test performance and failure rates that are consistent with the panelists' judgments. That set of combinations of test performance and failure rates can be represented by a line on a graph of percent correct scores versus percent failing. Suppose that the high and low values for percent correct scores for a panelist are represented by k_{max} and k_{min}, and the values for the percent failing are f_{max} and f_{min}. Then the endpoints of the line containing the set of acceptable values for the panelist are (k_{max}, f_{min}) and (k_{min}, f_{max}). The linear equation for the lines connecting those two points is

$$f = ak + b, \tag{8.4}$$

where

$$a = \frac{f_{min} - f_{max}}{k_{max} - k_{min}};$$

$$b = f_{min} - \frac{f_{min} - f_{max}}{k_{max} - k_{min}} k_{max}.$$

The value of a in Equation 8.4 is always negative because the numerator of the fraction is always negative, and the denominator is always positive. Thus, the line defined by Equation 8.4 has a negative slope. Suppose that $k_{max} = 55$, $k_{min} = 40$, $f_{max} = 35$ and $f_{min} = 25$, then $a = -\frac{2}{3}$ and $b = 61\frac{2}{3}$. This line is shown by the solid line in Figure 8.13.

The Hofstee (1983) method computes the estimate of the cut score by finding the intersection of the line defined in Equation 8.4 with the cumulative distribution function for the empirical data obtained from the test

$$f = F(k) \tag{8.5}$$

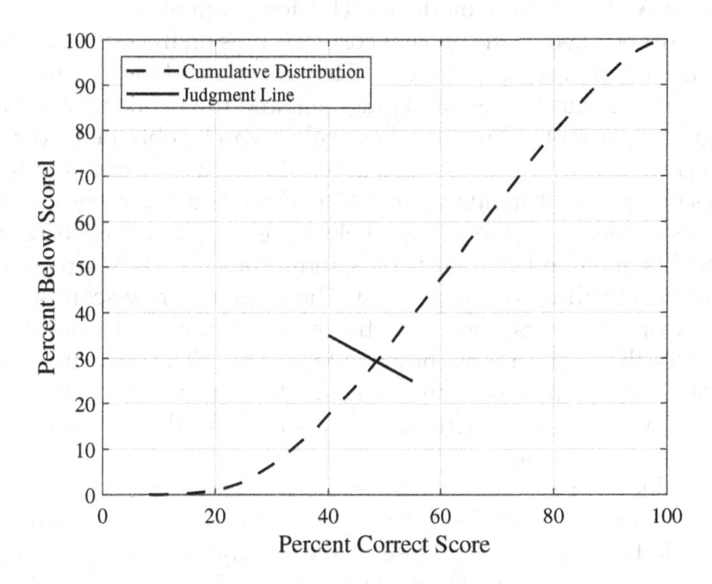

FIGURE 8.13
Cumulative score distribution and line based on panelist's judgment based on the Hofstee kernel.

where *f* and *k* are as defined in Equation 8.4 and *F*( ) represents the cumulative distribution function. The estimation of the cut score is found by solving Equations 8.4 and 8.5 as a set of simultaneous equations. Because $F(k)$ is an empirically derived function without a mathematical expression, the simultaneous equations are usually solved using graphic methods. A formal mathematical solution can be obtained if $F(k)$ is approximated by a mathematical expression. Figure 8.13 gives an example showing the intersection between the cumulative distribution function for the test scores (dashed line) and the line defined by Equation 8.4 (solid line).

A reasonable approach to modeling the error in the panelists' judgments when using the Hofstee (1983) kernel is not obvious. Yet it would be helpful to determine the influence of judgment error accuracy of the estimation of the standard on the reporting score scale. There are two different judgments required that use different scales—the percent correct scale as the reporting score scale and the percent failing scale as a means for including normative information into the standard setting process. The Hofstee kernel seems to require that panelists specify their own amount of judgment error on the two scales. But what is a reasonable magnitude for that error? It is also likely that those error estimates, as indicated by the width of the ranges that panelists specify, also have error.

As an initial attempt to model the judgment error for the Hofstee kernel, the example used for the contrasting groups method will be adapted to this kernel to allow for some comparisons. In particular, the interpretation of the policy definition and the reliability of judgments for 15 panelists that are provided in Table 8.1 will be used to model the Hofstee judgments. The values of k_{min} and k_{max} are specified as the percent correct scores on the test that operationally defines the reporting score scale. They correspond to the intended standard minus one standard error of judgment and the intended standard plus one standard error of judgment. These values can be obtained from the test characteristic curve for the test used in the standard setting study. Table 8.16 provides the intended standard, and the endpoints of the ranges for the percent correct scores and the percent below the standard that correspond to the panelists' intended standards plus and minus one standard error based on the rater reliability for the panelist. The estimated raw-score standard is obtained from the intersection of each panelist's line and the cumulative distribution for the test scores as shown in Figure 8.13. The estimated values on the θ-scale are interpolated from the estimated raw score and the test characteristic curve. The same item parameters used for the contracting groups example were used here.

The results from this simulation of the translation process for panelists using the Hofstee kernel are very good. The estimated standards on the scale implied by $\mathbf{D}_P$ are very close to the interpretations of the standards by the simulated panelists. However, this is an ideal case where the panelists accurately identified score ranges and percent failing ranges that were consistent with their intended standard. It is unlikely that panelists' actual

TABLE 8.16

Panelists' Standard, Percent Correct Range, Percent-Below Range, and Estimated Standard Using the Hofstee Kernel

Panelist	Panelist Interpretation of Standard	k_{min}	k_{max}	f_{min}	f_{max}	Estimated Standard Proportion Correct	Estimated Standard θ-Scale
1	–1.09	29	52	7	20	38	–1.08
2	–1.06	29	54	8	21	38	–1.08
3	–1.09	30	50	7	20	38	–1.08
4	–1.09	29	52	7	20	38	–1.08
5	–1.00	29	56	9	23	40	–0.99
6	–0.92	32	54	10	26	42	–0.90
7	–1.09	29	52	7	20	38	–1.08
8	–0.98	30	55	9	24	40	–0.99
9	–0.98	31	54	9	24	40	–0.99
10	–1.09	28	55	7	21	38	–1.08
11	–1.01	30	55	9	23	40	–0.99
12	–0.95	31	56	10	25	42	–0.90
13	–1.04	30	53	8	22	40	–0.99
14	–0.92	30	57	10	25	42	–0.90
15	–1.09	29	52	7	21	38	–1.08
Mean	–1.03	29.76	53.76	8.20	22.38	39.56	–1.02

values for these ranges would be this accurate. To check the sensitivity of the Hofstee kernel to error in specification of the endpoints, the endpoints of the ranges were randomly sampled from error distributions that were consistent with the reliability of the judgments used for the contrasting groups example. For the proportion correct scores, k_{min} and k_{max}, the endpoints of the range were obtained by adding and subtracting a randomly selected standard error value from a chi-square distribution with 7 degrees of freedom that was scaled to have a mean equal to the standard error computed from the reliability for each panelist. Comparing the mean values from Tables 8.16 and 8.17 shows that the mean range was about the same with and without the added error. When error was included, the lower bound of the range was slightly lower: 27.38 versus 29.76.

Error was added to the judgments of the estimates of the range of percent below the cut score by sampling from a beta distribution that had a mean equal to the proportion below given in Table 8.16 and the same process for computing the variation in the proportions used for the Angoff probability-estimation method described earlier in this chapter. The mean range with and without error is similar: 14.37% versus 14.18% differences. However, the addition of error results in an increase of about 1 percentage point at each end of the range.

TABLE 8.17

Panelists' Standard, Percent Correct Range, Percent-Below Range, and Estimated Standard Using the Hofstee Kernel for Ratings with Error

Panelist	Panelist Interpretation of Standard	k_{min}	k_{max}	f_{min}	f_{max}	Estimated Standard Proportion Correct	Estimated Standard θ-Scale
1	−1.09	29	54	4	7	25.00	−2.27
2	−1.06	26	57	2	26	38.33	−1.08
3	−1.09	30	48	2	20	35.00	−1.29
4	−1.09	22	57	6	22	33.33	−1.40
5	−1.00	19	56	25	28	23.33	−2.58
6	−0.92	34	55	17	30	35.00	−1.29
7	−1.09	26	53	3	13	31.67	−1.53
8	−0.98	23	57	26	29	25.00	−2.27
9	−0.98	33	54	13	17	28.33	−1.84
10	−1.09	19	49	5	13	28.33	−1.84
11	−1.01	31	59	12	16	28.33	−1.84
12	−0.95	33	53	5	31	40.00	−0.99
13	−1.04	24	51	3	28	36.67	−1.18
14	−0.92	34	54	9	34	40.00	−0.99
15	−1.09	27	52	4	38	40.00	−0.99
Mean	−1.03	27.38	53.91	9.05	23.42	32.56	−1.56

Despite these slight differences in mean values for the estimates of the endpoints of the ranges, the standard estimated using the Hofstee kernel is quite different when errors in judgments are present in the data used for computing the panelists' estimates. The mean percent correct score value that corresponds to the standard is 7 percentage points lower when error is included than when it is not. That difference corresponds to a .54 difference on the θ-scale—about half a standard deviation for the distribution of examinees. The major contributors to that difference are the panelists that had errors that resulted in small ranges of values for the percent below the standard. Those small ranges result in small values for the slope of the line that intersects with the cumulative distribution of scores. When the slope is low, the line intersects the cumulative distribution at a lower point than when the slope is high. The results in this case are very low estimates of the standard as is shown by the estimates for Panelists 1, 5, and 8.

While it is clear that some of the panelists' estimates do not accurately recover the intended standard and that the mean of the estimates from the 15 simulated panelists underestimates the intended standard by a substantial amount, that is not the case when the Hofstee estimation process is used with the means of the separate panelist judgments. Figure 8.14 shows the intersection of the line defined by the means of the panelists' judgments and the cumulative distribution function for the test. The intersection of the two

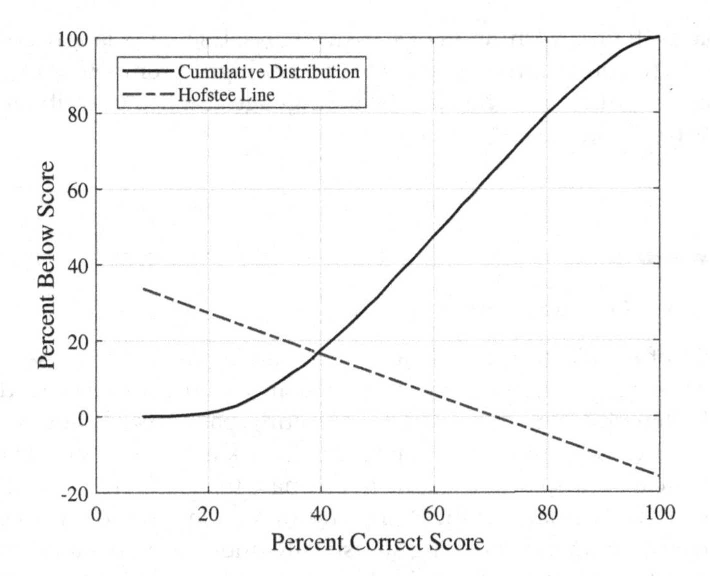

FIGURE 8.14
Cumulative score distribution and line based on mean of panelist's judgment based on the Hofstee kernel.

lines is at approximately 40 on the percent correct scale. When that value is converted to the θ-scale, the estimated standard is −.99, very close to the intended standard. Thus, for the kinds of panelists' judgment processes and judgment errors modeled in the example, using the mean values of the judgments to determine the intersection of the two lines yields a good estimate of the standard. Unfortunately, the relationship of this modeling of panelists to what they do in practice is unknown. In this simulation, errors in judgment were symmetric around the intended standard—statistical bias was not included in the simulation. But panelists might have an aversion to specifying values that they perceive as too high or too low. For example, the values for $f_{\min}$ in Table 8.17 are often in single digits. Panelists might consider percent of failure in that range might not be credible, even for a minimum competency test. That might result in judgments by the panelists that had a positive bias for the lower bound of the range. The opposite effect could be present for standards that had a very high failure rate such as ones used for selecting scholarship candidates. Such influences would tend to bias the estimation of the standard toward the middle of the reporting score scale.

This limited example indicates the importance of selecting panelists for the standard setting process that are highly knowledgeable about the relationship between performance on the test and interpretations made from that level of performance, and about the characteristics of the examinee sample that is employed to produce the cumulative distribution function used to estimate the standard. Panelists without the appropriate background will likely produce judgments from the Hofstee kernel that are very unstable with unknown types of statistical bias. Using the language translation analogy, translators are needed who are highly competent in both the language of policy and the statistical language used to describe tests and score distributions.

8.4 Recent Developments

The field of standard setting is a very dynamic one. Each of the common standard setting kernels is implemented in practice in slightly different ways. Different criteria are used for selecting panelists, different training processes are used, and many types of feedback are provided. Also, different statistical approaches are used to estimate the standards on the reporting score scale—usually without any research to determine if the estimates meet any requirements for being statistically unbiased with small error. It is impossible for any study of standard setting to address all the possible variations that occur in practice.

There are certain trends in the field that suggest that developers and practitioners are working to tighten the connections between the component parts of a standard setting process. For example, Ferrara, Svetina, Skucha, and Davidson (2011) are working to tighten the connections between $\mathbf{D}_C$ and the test that is used to operationally define the construct implied by $\mathbf{D}_P$. They suggest that the $\mathbf{D}_C$ should be written first and items should be written to match the requirements of the detailed content descriptions. They list a requirement that "items are located on the within-grade test scales so that item response demands are aligned with corresponding proficiency level descriptors" (p. 3). They are optimistic that "we can learn how to train item writers to develop items that hit difficulty range targets and targeted proficiency levels" (p. 4).

Lewis and Cook (2020) assume the type of test construction suggested by Ferrara et al. (2011) to propose a new standard setting kernel labeled "embedded standard setting" (ESS). To determine a cut score using the ESS method, items are written to match the requirements of $\mathbf{D}_C$ and then they are calibrated using an IRT model. Then the items are ordered according to a mapping criterion. Lewis and Cook (2020) use the traditional bookmark mapping criterion of .67, but they admit that the mapping criterion makes a difference. The mapping criterion is the definition of what it means to "know" an item in an achievement level. They suggest that the mapping criterion might be the value that minimizes inconsistencies in classifications of items relative to the standard. Items need to be produced for at least two levels—above and below the cut score. The cut score is determined by finding the item location that minimizes the misclassification of items (the inconsistencies) into categories above and below the cut score based on the target the item writers had for the item. The cut score is the mapped point on the scale for the item that minimizes the inconsistencies. This kernel assigns the translation process to the persons writing the test items rather than a recruited panel. The ESS kernel makes strong assumptions about item writers' understanding of $\mathbf{D_P}$ and $\mathbf{D_C}$ and their capabilities to produce well-targeted test items.

Other current trends in the standard setting literature are described by Peabody and Wind (2019). One of the trends is to do standard setting online without the requirement that panelists meet at a single location. Computer and internet capabilities now make remote sessions a practical possibility. This type of standard setting process allows more accurate replication of the process because much of the training can be recorded. Also, it is possible to collect more information from each panelist about the process that they use when making judgments about test items related to $\mathbf{D_P}$ and $\mathbf{D_C}$. Peabody and Wind (2019) use these capabilities to investigate the relationship between panelists' knowledge of the content of the test and the level of standard that results from the ratings they provide. Their research shows that there is a relationship and panelists with higher levels of proficiency tend to set higher standards. The implication is that the way that panelists are chosen is very important and that it may be advisable to use information about panelists' levels of proficiency when estimating standards.

An unfortunate aspect of work on the development and implementation of standard setting procedures is that much of it is not available in archival publications. Many applications are performed as part of contract work for a specific testing program and the details of the process may only be communicated to the organization calling for the standard setting study. Reports of standard setting studies often lack details about the selection of panelists, the training process, the level of comprehension of panelists about their tasks, etc. It would be extremely helpful for advancing the research on standard setting if published articles would have links to materials used in standard setting studies and detailed reports of those studies.

8.5 Cautions about Generalizability

The simulation results presented for the various standard setting kernels and their variations are based on the results of real studies when possible, or on the author's judgments of how panelists might function when participating in the tasks specified by a standard setting kernel. The specifications for reliability of judgments of panelists and the error distributions for those judgments are selected to be consistent with the research literature, but only a limited number of the large number of possibilities were used for the examples in this chapter. A thorough report of the functioning of estimation procedures used with standard setting kernels would require another book-length document. The examples used here are meant to be thought provoking and to present an approach to checking the functioning of standard setting kernels, but they are far from definitive. They assume that the simulation processes approximate the way panelists approach the judgment tasks. The simulation approaches were selected to approximate reality, but standard setting processes are so varied that it is unlikely that these simulations match panelists' behavior in more than a small percentage of standard setting studies. The level of match will depend on the way the tasks for panelists are described to them, the amount of training they receive, their levels of understanding of the tests, and many other design features for the standard setting process. It is important that those who design and implement standard setting processes perform careful simulation studies to investigate their own implementations of standard setting kernels. It is also important that more research be done into the ways panelists perform the standard setting tasks. There are a few "think aloud" or "cognitive laboratory" studies that give some insights into the cognitive processes of panelists, but these are rare.

8.6 Summary and Conclusion

There are several important findings reported in this chapter. First, the statistical methods that are used to estimate the location of standards on the reporting score scales for a test are not equally good at recovering the standard intended by the agency that calls for the standard. As with all translations from one language to another, the results can range from inaccurate translations that may be misleading to highly nuanced translations that capture the details of an author's intent. I think back to my first experience with having a technical article on computerized adaptive testing that was written in German translated into English by a native speaker of German with little knowledge of mathematics or statistics. Words with well-defined mathematical meaning like "set" have multiple common usage words that can be selected by a translator such as "group" or "array," which have their own specific mathematical meanings. I also remember giving a presentation to a group in China with a simultaneous translation when I used the term "item pool" and had it translated to the Mandarin equivalent of "item ocean" by my translator. The nuances of the translations are important.

An example of the differences in quality of translations given here are the estimates provided by the Angoff probability kernel and the Angoff item-score kernel. Under the error assumptions used in the simulations reported in this chapter, the translations to the number-correct score scale for the two methods are quite different. The argument made in this chapter is that the two standard setting kernels are providing data that are conceptually different and that the data do not provide equally good estimates of the intended standard. Examples of challenges to recovering the intended standard were shown for other standard setting kernels as well.

The second important finding is that a statistical and/or psychometric analysis of the information provided by a standard setting kernel can give guidance about how to improve the estimate of the intended standard. When the data from the Angoff item-score kernel is mapped through the proper function (not the test characteristic curve), it can recover the intended standard with reasonable accuracy. Another good example is shown with different ways of estimating the standard using the Hofstee kernel. Using the average of the estimated standards from each of the panelists gives very poor estimates of the intended standard, but using the average of each of the estimates of the endpoints of the intervals to get the standard gives reasonable recovery of the intended standard. These results indicate that it is important that those who design standard setting studies do not arbitrarily select a method for converting the panelists' judgments to the reporting score scale; they should perform careful statistical analyses of the conversion method to determine if it supports the recovery of the intended standard.

A third important finding implied by this chapter is that little is known about the distributions of errors of judgments made by panelists with different characteristics. There have been some research studies, such as Peabody and Wind (2019), that show that panelists with different characteristics produce different judgments. The cognitive processes that result in those different judgments are not known. As a result, it is difficult to decide on the amount of confidence to place in the simulation studies used in this chapter. A positive feature of doing those simulations is that the research questions about panelist rating processes become very evident. One reason for writing this book is to stimulate more research about how the panelists perform their tasks. This is difficult research to perform. The potential gains from that research are very great.

9

Evaluating the Standard Setting Process

9.1 Introduction

The position taken in this book is that standard setting is a process for translating the conceptual description of a standard ($\mathbf{D_P}$) produced by an agency (Ag) that calls for the standard into the language of the reporting score scale for a test that is designed to measure the construct (ζ) implied by that same conceptual description. An important question after completing a standard setting process is whether the estimated cut score ($\hat{\zeta}_{Ag}$) on the reporting score scale accurately reflects the intentions of the agency (ζ_{Ag}) when they produced the policy calling for the standard. Any translation task can be done with a range of quality from poorly to extremely well. And even two equally good translations of the same text may have differences. A challenge when evaluating the quality of the outcomes of a standard setting study is determining whether observed variations in the judgments of panelists are the expected variation from excellent translations or are due to misunderstandings and mistranslations of key ideas.

Current thinking on the evaluation of the results of a standard setting process is based on the work of Kane (1994). That work emphasized that users make inferences about examinees from the results of a standard setting study, and it is those inferences that need to be supported by evidence. The structure of the support for the inferences is in the form of what Kane (1994) called a validity argument. This argument is a series of statements, often called claims, that lead from the policy definition $\mathbf{D_P}$ to the inferences made from the scores on the test relative to the estimate of the cut score. Each of the statements needs to be supported by evidence. The judged quality of the results of a standard setting study depends on the quality of the evidence collected to support each statement in the validity argument.

The position taken by Kane (1994) was the result of evolution in thinking about how to show that the results of a standard setting process is credible. That position developed over many years. Linn, Madaus, and Pedulla (1982) provided an early discussion of some of the components of a validity

DOI: 10.1201/9780429156410-9

argument in the context of minimum competency testing of students within the educational system.

> When performance on an MCT [minimum competency test] is used to make a decision about certifying the completion of a given level of education, the inference made when a pupil is retained in grade or denied a high school diploma is that the pupil has not mastered a minimum body of material appropriate for that level of education.
>
> **(p. 20)**

They also indicated that often there is little construct validity evidence for the test used in the standard setting process. Nor is there research done to check if those classified below or above a cut score are different in their performance on other criteria. They stated the desired inference and described some of the evidence needed to support a validity argument for that inference.

Jaeger (1989) made an even more explicit connection to the idea of a validity argument in the context of educational testing:

> Such classification [above a cutscore] is an *inference* about the student's status with respect to a construct labeled *competence*, as well as a prediction about the student's likely performance in a higher level grade, in a postsecondary education setting, or in adult society. When tests are used to make inferences such as these, examination of the validity of the inferences is mandatory.
>
> **(p. 487)**

Kane (1994) emphasized that validity refers to an interpretation (inference) made from the test scores. The validity of a test, the scores on the test, or classifications based on the scores cannot be evaluated until the interpretation (inference) is stated. Then the interpretation is supported (in informal use labeled as validated) if there is appropriate evidence. It is important to specify the details of the interpretation so that each claim is clear. Hambleton (1998) also indicated that the importance of the decision should guide how much evidence is needed. Evidence needs to be collected about each claim to support the validity of the interpretation. That evidence is specific to each inference made about a particular population and a specified context.

Kane (1994, 2001) suggested organizing evidence related to the inferences made from a standard setting process into conceptual, procedural, internal, and external categories. He indicated that the validity investigations should be iterative so that consistency between the various parts of the argument could be checked and maintained. The connections between parts of the validity argument are called an interpretive argument. The interpretive argument gives a framework for the information needed to support the validity of the inferences. Later work (Clauser, Kane, Clauser, 2020) refined the list of four sources of evidence for the credibility of a standard setting—conceptual coherence, procedural evidence, evidence of internal consistency, and evidence based on external criteria. They also emphasized the replicability of the standard over panels and investigations of the sources of error related to replicability. Pitoniak and Cizek (2016) provided an interpretation of three of Kane's categories: procedural, internal, and external. In their view, procedural refers to whether the procedures were conducted as planned; internal is whether the results are replicable and consistent; and external is whether the results agree with other information external to the entire process. They stated that a validity report including evidence for each of the categories should be completed after every standard setting study.

Kane (1994, p. 457) provided some necessary cautions about expectations for the conclusions from evidence for the validity argument. He stated that "even if all available checks on the validity of the standard are implemented, the best that researchers are likely to be able to do is to show that the proposed standard is reasonable." His statement is consistent with the observation that expert translators may give slightly different translations of a text. That does not mean that one is correct and the other one is not. It more likely means that the connection between one language representation and that in a different language is not a one-to-one function. Similarly, the translation from a policy definition to a point on a reporting score scale is not likely a one-to-one function. The goal of the evaluation of a standard setting process is to determine if the results are "reasonable" as stated by Kane (1994).

9.2 Types of Evidence

This section of the chapter is structured according to Kane's (1994) organization of types of evidence. These categories overlap and evidence can often be classified as more than one type. However, his categories do give some structure to the requirements needed to support the use of an estimated standard. Those requirements may seem overwhelming. The structure helps divide the requirements into manageable pieces.

9.2.1 Conceptual Coherence Evidence

Conceptual coherence evidence is the most amorphous of Kane's (1994) categories. This type of evidence requires those who design a standard setting process to demonstrate that they understand the "big picture" related to setting standards rather than having put together a string of procedures. Cizek and Bunch (2007) put this in broad terms when they discuss the requirements given in the 1999 edition of the Standards for Educational and Psychological Testing (AERA, APA & NCME, 1999). Their summary of Standard 4.19 states that reports of a standard setting process should document "the rationale and procedures used for establishing cut scores" (p. 58). Of course, what is needed to document a rationale is somewhat vague. In a sense, everything in this book is documenting the rationale for a standard setting process.

As a beginning point for conceptual coherence, Kane (1994) made a specific distinction between the passing score (ζ_{Ag}, the point on the continuum defined by the agency) and the performance standard ($\mathbf{D}_P$). He also made an important distinction between the match of the estimated passing score ($\hat{\zeta}_{Ag}$) to the performance standard and the reasonableness of the performance standard. This distinction divides the validity argument into two parts. The first part is whether the performance standard is appropriate? This is a policy issue. He called this the policy assumption (p. 456). The second part is whether the passing score is consistent with the performance standard?

> The best that we can do in supporting the choice of a performance standard and an associated passing score is to show that the passing score is consistent with the proposed performance standard and that this standard of performance represents a reasonable choice, given the overall goals of the assessment program. (p. 437)

He called this the descriptive assumption (p. 456). He indicated the second part of the argument, supporting the match between the cut score and the policy definition, is much more manageable than the first part, supporting that the policy definition is reasonable. Sometimes, critics question the reasonableness of the estimated standard on the reporting score scale when they really disagree with the policy. The estimated standard may be an accurate translation of $\mathbf{D}_P$, but some may question the reasonableness of $\mathbf{D}_P$. This is a source of confusion.

Cizek and Bunch (2007) indicated that there may be other challenges to supporting the results of the standard setting process because there may be multiple inferences made from the estimated standard. In a sense, there may be multiple policy definitions calling for the standard even though they are presented as a single $\mathbf{D}_P$. The standard set on a licensure examination may be the translation of a policy about the required skills and knowledge for entry into a profession, but it may also support policy used to control the market for persons in the profession or policy to increase the prestige of a field. Standards on an educational test may be translations of policy concerning teacher qualifications and expectations for outcomes of educational programs as well as policy about goals for students' learning of the content and skills specified by curriculum.

The practical implications of the requirements from the *Standards* (1999) and Kane's (1994) recommendation for a formal validity argument to support the interpretations desired from the use of the standards are that there should be:

1. A clear statement that indicates the number of outcomes intended by the policy;

2. Documentation of the connection between the $\mathbf{D_P}$ and the $\mathbf{D_C}$ that is the target for the test;
3. Documentation of the relationship of the test that operationalizes the continuum with both the $\mathbf{D_P}$ and $\mathbf{D_C}$;
4. An explanation for why the information gained from a selected standard setting kernel will be sufficient for accurately translating the intentions in the $\mathbf{D_P}$ to the score scale defined by the test.

Most of this part of the validity argument is conceptual and the support for it comes from a careful review of the descriptions of the connections between the different component parts of the argument. Some of these descriptions are in the form of claims about the relationships between the parts. For example, the $\mathbf{D_P}$ may include a claim that it is reasonable to order individuals along a continuum of proficiency and this continuum may be a composite of the skills and knowledge described in the $\mathbf{D_C}$. Further, there may be another stated or implied claim that the score scale defined by the test used in the standard setting process gives a good approximation to the target proficiency continuum.

9.2.1.1 The Policy Definition (D_P)

The policy definition of the standard ($\mathbf{D_P}$) may be considered to have strong supporting evidence by the fact that the agency that calls for the standard produced it. Persons may disagree with the position taken by the agency, but it is their standard, so it is difficult to argue that it is incorrect. At most, it might be argued, as suggested by Kane (1994), that it is unreasonable. But the way that $\mathbf{D_P}$ describes the standard is especially important. As Shepard (1980) indicated: "The validity of judgmentally set standards depends on the definition of the minimally qualified examinee" (p. 452). If the agency does not produce a precise, clearly stated definition, translators will have difficulty transforming the agency's intentions into the language of the reporting score scale for the test. When evaluating the standard setting process, it makes sense to begin with the clarity and precision of the starting point of the process, $\mathbf{D_P}$.

Despite the importance of the $\mathbf{D_P}$ for setting a standard, little emphasis has been given to evaluation of the usefulness of $\mathbf{D_P}$ as the basis for the translation to a point on the reporting score scale. Those who typically design and conduct standard setting studies are not in a good position to evaluate $\mathbf{D_P}$ because the agency hires them. In some cases, there are external reviews of the entire process, but those seldom give thoughtful consideration to the language of $\mathbf{D_P}$. For example, the review of standard setting on the NAEP Science assessment performed by a committee empaneled by the National Research Council (Pellegrino, Jones & Mitchell, 1999) extensively reviewed the reporting of results of NAEP using the standard set on the reporting score scale but paid only minor attention to the actual language of the policy. They state:

> The public may be misled if they infer a different meaning from the achievement level descriptions than is intended. (For example, for performance at the advanced level, the public and policy makers could infer a meaning based on other uses of the label "advanced," such as advanced placement, which implies a different standard. ...) (pp. 164–165)

One of the review committee's conclusions was as follows:

> *Conclusion 5A.* Standards-based reporting is intended to be useful in communicating student results to the public and policy makers. However, sufficient research is not yet available to determine how various audiences interpret and use NAEP's achievement-level results.
>
> **(p. 182)**

Some work has been done in the last 20 years to investigate the differences in meaning of words used in $\mathbf{D_P}$, but it is difficult to determine if this work has had any influence on the language used by an agency to describe a standard. Burt and Stapleton (2010) did a research study on the differences in implied meanings of words typically used to describe standards. Their results showed that words like "satisfactory" and "proficient" have significant differences in connotation of levels of mastery. They also reported the wide variety of phrases states use for similar levels of performance when reporting state assessment results in the United States. These implied differences in meaning of these phrases are reduced when the agency provides a definition of what they mean by a specific word or phrase, but the differences may still be highly significant. Word choice and the amount of elaboration used to describe standards are important when communicating the intentions of the agency. Burt and Stapleton (2010) conclude:

> that some performance-level labels may elicit perceptions of differential levels of knowledge: these differences could continue through the standard setting sessions and penetrate to the cut scores, but that is less likely if the standard-setting method utilizes definitions at the beginning of the standard setting process. (p. 36).

Given the evidence that selection of words and phrases used in $\mathbf{D_P}$ has important implications for the process of translating the intentions of the agency into a point on the reporting score scale, it is important that $\mathbf{D_P}$ be reviewed against recommendations for good practice. Perie (2008) provides an excellent guide to use for this review. For the case of policy definitions related to standards for the outcomes of educational programs, she recommends:

> The definitions should clearly state the degree of knowledge and skills expected of students at each performance level. They should be concise, approximately 1–2 sentences, and clear. Because it is the backbone of all further writing, policymakers should carefully consider the wording and be sure each definition communicates the intended goals and clearly distinguishes one level from the next.
>
> **(p. 18)**

Perie (2008) provides some specific examples of $\mathbf{D_P}$ from different state programs with commentary about the qualities of the different definitions. These examples can be helpful when evaluating the $\mathbf{D_P}$ connected with a standard setting activity and can provide support for the beginning part of the validity argument.

9.2.1.2 The Content-Specific Definition (D_C)

Although there is no formal script for creating a validity argument to support the results of a standard setting process, the usual first claim is that $\mathbf{D_P}$ represents the standard intended by the agency and a common second claim is that the content-specific definition of knowledge, skills, and abilities, $\mathbf{D_C}$, that corresponds to the standard is consistent with $\mathbf{D_P}$ and the agency's intentions. Early standard setting processes often had panelists draft $\mathbf{D_C}$ during the standard setting process. It was an activity designed to help the panelists understand the knowledge, skills, and abilities of those persons who are above the standard. A second part of that process was to specify $\mathbf{minD_C}$ as another activity to help panelists come to a common understanding of what they were to be translating using the tasks in the standard setting kernel. A problem with having the panelists produce $\mathbf{D_C}$ during the standard setting process is that the agency does not have an opportunity to judge whether $\mathbf{D_C}$ and $\mathbf{minD_C}$ are consistent with their intentions until the standard setting process is complete. If these definitions are not consistent with each other, the results of the standard setting process are open to question. Shepard, Glaser, Linn, and Bohrnstedt (1993) were very critical of producing drafts of $\mathbf{D_C}$ and $\mathbf{minD_C}$ during the standard setting process for NAEP. To address this issue, later standard setting process used for NAEP (Loomis & Bourque, 2001) included a process where $\mathbf{D_C}$ was created by a panel of selected content experts prior to the beginning of the standard setting process. This allows the agency to review and edit $\mathbf{D_C}$ before the standard setting process to ensure consistency with $\mathbf{D_P}$ and their intentions. $\mathbf{D_C}$ was provided to panelists participating in the standard setting process along with $\mathbf{D_P}$ with context information that indicated the agency had approved both definitions. This approach supports a strong connection between definitions for the validity argument for the standard. Of course, good documentation is needed of the review process and statements of endorsement by the agency.

Because of the importance of $\mathbf{D_C}$ to the standard setting process, there is now specific guidance about how the definitions should be produced and the level of detail required. Egan, Schneider, and Ferrara (2012) have provided a chapter that gives a framework for what they call "performance level descriptors" (PLDs). Perie (2008) also gave detailed guidance for producing PLDs that are tightly connected to $\mathbf{D_P}$. It is important that $\mathbf{D_C}$ (i.e., PLDs) have sufficient detail so that connection to the specifications for the test that is the operational definition of the construct implied by $\mathbf{D_P}$ can be determined with a level of confidence. $\mathbf{D_C}$ is a general description of the construct for the specific subject matter area, while the test specifications are the more detailed description of the construct.

9.2.1.3 The Operational Definition of the Intended Construct, ζ

When the agency calls for one or more standards and produces the policy definition, $\mathbf{D_P}$, they are implying a continuum of performance and the standard is the location of a point on the continuum. In the psychometric literature, this continuum is often labeled as a construct because it is an abstract concept that is "constructed" in the minds of those who comprise the agency. The test used in the standard setting process provides the reporting score scale that is the operational definition of the construct. An important part of the validity argument is providing evidence that the reporting score scale for the test is a good representation of the construct implied by $\mathbf{D_C}$.

There are subtleties to the concept of the continuum and the ways that tests are designed to provide measures of constructs. Haertel (1985) made a distinction between the types of constructs hypothesized in the psychology literature and those used to define the scale for tests of achievement and proficiency. He argued that achievement constructs are less stable than personality or aptitude constructs because they are defined by goals of instruction. There is an expectation that individuals will change their locations on an achievement construct with instruction. Achievement constructs are also more narrowly defined than other psychological constructs because they are directly tied to content and process expectations and the desired products of instruction. This makes generalizability beyond the narrowly defined content domain difficult. If there is a desire for such generalization, Haertel (1985) argues that "content validity alone is insufficient" (p. 42). The issue of generalizability may not be a problem for lengthy certification/licensure tests designed based on extensive job analyses, but generalization may be an issue for assessments like NAEP that are attempting to make inferences about the functioning of the diverse educational systems in the United States.

Since that early discussion of the characteristics of the construct/continuum underlying the standard setting process, there has been increasing emphasis on the connection between $\mathbf{D_P}$ and $\mathbf{D_C}$ and the design and construction of the test that is the operational definition of the construct. Ferrara, Svetina, Skucha, and Davidson (2011) indicated that content standards (i.e.,

$\mathbf{D_C}$) and construct definition implied by $\mathbf{D_P}$ should be part of the test design process. They suggested that the $\mathbf{D_C}$ (what they label as achievement-level descriptions [ALDs]) should be written before the test specifications and items should be written to match the requirements of the ALDs. They indicated how they believe test items should be developed to support a standard setting process. They believe that for a well-designed test, "items are located on the within-grade test scales so that item response demands are aligned with corresponding proficiency level descriptors" (p. 3). They admit that in the past, test items were not written to specifically match the requirements of standard setting but are optimistic that "we can learn how to train item writers to develop items that hit difficulty range targets and targeted proficiency levels" (p. 4).

Everson and Forte (2020) elaborate on the recommendations made by Ferrara, Svetina, Skucha, and Davidson (2011) and make the connection between the standard setting process and test development process very explicit.

> Write achievement level descriptors (ALDs) that provide clear and detailed statements about the nature of content-based performance progressions, use those ALDs to write items that align with the score scale from the outset, and listen to what the performance data say about cut scores and alignment. (p. 7)

The positions taken by these authors indicate that the validity argument for the inferences made from a standard must include evidence to support a claim that the reporting score scale for the test is a good representation of the construct implied by $\mathbf{D_P}$. This is not a new idea. Shepard, Glaser, Linn, and Bohrnstedt (1993) reported evidence of that type as part of an evaluation of the standard setting process conducted on NAEP. The evaluation team convened a validity panel that had as part of their agenda the review of the connection between policy and content descriptions, and the test items selected as exemplars. Shepard et al. (1993) did not phrase the question to the validity panel as the connection between the intended and operational constructs. The question they posed was whether examinees classified as above a standard could do the things listed in the content definition $\mathbf{D_C}$. That question is much broader than one about the constructs. It is the major claim for the full validity argument.

The current approach to the issue of the connection between the definitions is to provide evidence about the test design and development process and then to provide evidence from alignment studies to show that the design and development process was successful. For a certification/licensure test, the test design and development processes usually begin with a job analysis to determine the skills and knowledge required by a successful candidate. Then that information is used to develop the test specifications followed by item development specifications for the creation of items for an item pool.

After test items are written to meet the requirements of the item development specifications, they are screened for quality and items are selected to match the test specifications from those that pass the screening. Finally, an alignment study can be performed to get evidence about the level of success of the process to construct a test that matches the specifications.

A persuasive argument for the connection between the desired construct and the operational score scale can be built on support for the following claims:

1. The job analysis provides an accurate depiction of the knowledge, skills, and abilities required by the occupation that is the focus of the standard setting.
2. The test specifications describe the characteristics of a test that provides an adequate sampling of the knowledge, skills, and abilities needed to make the desired inferences about performance in the profession.
3. The item development process produces items that elicit observable responses from examinees that can be used to assess their levels of performance related to the desired knowledge, skills, and abilities.
4. The test form construction process results in the selection of test items that match the requirements of the test specifications (alignment).
5. The test is sufficiently reliable/precise to locate individuals on the continuum implied by the test specifications.

The test that is the focus of standard setting in the context of education has similar requirements. The major difference is in Claim 1 on the list. Rather than basing the content of the test on a job analysis, the starting point is a description of the curriculum that is the focus of instruction. Part of the description of the curriculum should be a description of expectations for the knowledge, skills, and abilities that students are expected to acquire. Those expectations are the basis for the test specifications given in Claim 2.

Figure 9.1 gives a graphic representation between these initial components of the validity arguments. The stronger the evidence for the connections represented by the arrows, the stronger the support for the inferences that are desired from the standard setting process.

9.2.2 Procedural Evidence

Procedural evidence is the information provided to show that the process proposed for collecting information for the estimation of the standard was conducted properly. There are three aspects to procedural evidence: (1) information about the selection of the kernel and design for the standard setting process, (2) indications that panelists understand and can perform their tasks, and (3) documentation of the standard setting process. Published

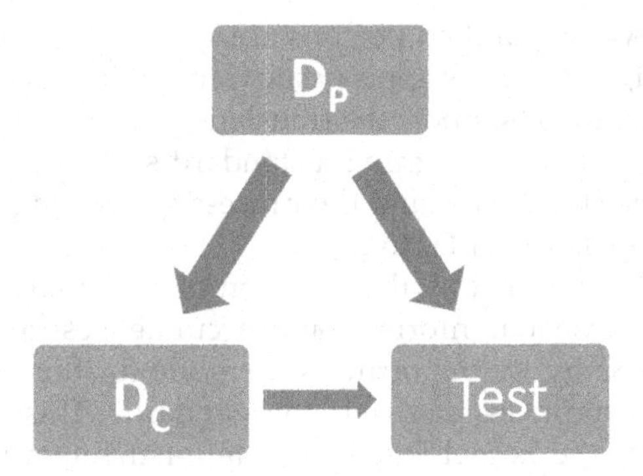

FIGURE 9.1
Relationships among the initial components of the validity argument.

reports of standard setting studies tend to provide substantial information about the second element in this list—panelist selection and panelists' self-reports of their understanding of the tasks they were asked to do. There is less information provided about the rationale for the selection of the kernel and the design of the study.

9.2.2.1 Kernel Selection and Design of the Process

Reports of standard setting processes typically have little or no information about why a particular kernel was selected as the basis for the process. The exception is when the test used to operationally define the continuum for the standard contains several different item types, such as multiple-choice and extended-response items. Then a rationale is provided for the use of more than one kernel in the process. A kernel should be selected that is consistent with the requirements of each item type.

Although little information is typically given about the rationale for selecting a specific kernel for a standard setting process, the results of a standard setting process are sometimes questioned because of the selection of the kernel. For example, Shepard, Glaser, Linn, and Bohrnstedt (1993) believed that the Angoff probability estimation kernel was fatally flawed because they did not think panelists were capable of accurately making conditional *p*-value estimates (p. 72). They also believed that setting standards based on item-by-item ratings deprived panelists of the full context of examinee performance. Pearson and DeStephano (1993) concurred with that view. They stated: "It may be that the demand characteristics of a process that requires the assignment of ratings by imagining the *actual* performance of a *hypothetical* group based on *ideal* achievement is beyond the cognitive capacity of even the most knowledgeable judges" (p. 155).

The criticisms of standard setting kernels may be overstated. It may be that the task given to panelists is very challenging—the translation of any text from one language to another is challenging. But that does not mean that it cannot be done when qualified persons are selected for the task, the task is clearly described, and there is appropriate training in the methodology used to collect information to support the translation. On the other hand, it is the obligation of the persons designing a standard setting process to give evidence that the selected kernel and the process supporting its use can result in an accurate translation of $\mathbf{D}_P$ to ζ_{Ag}.

The claim that is implied by the selection of a standard setting kernel is that its use gives sufficient information to accurately estimate a point on the continuum defined by the reporting score scale for the test used to make decisions about examinees relative to the standards. This claim has embedded within it that the kernel is appropriate for the item type used to collect information about examinee's capabilities and the scoring model used to estimate the examinees' locations on the score scale. If the test used to estimate the location of examinees has open-ended items scored using rating scales, then the kernel needs to support translation to the score scale for tests composed of such test items. If more than one item type is used, it may be necessary to have multiple standard setting kernels to acquire the information needed for the translation of $\mathbf{D}_P$ to ζ_{Ag}.

Along with a logical argument that the selected kernel is appropriate for the characteristics of the test used to operationally define the continuum for the standard, it is also useful to have evidence that the information obtained using the kernel can be used to recover the standard intended by the agency. Chapters 5 and 8 in this book give examples of approaches for documenting the quality of estimates of standards obtained from the use of a kernel. These indicators of quality may include showing that the estimates of the standard are statistically unbiased and have small variance. Another consideration is what van der Linden (1995) calls "feasibility"—is it possible to obtain the value of the intended standard? In some cases, the standard setting kernel will support a limited number of values for the standard. If a test is composed of a single essay evaluated on a 5-point rubric, and the standard setting kernel requires panelists to indicate the score on the essay the minimally qualified individual will most likely obtain, then there are only five values that can be recommended as the translation of the standard. It may not be possible for a panelist to provide information using that kernel that will accurately translate to the intended standard. However, another kernel, such as the panelist estimates the mean score on the essay for 100 examinees described by $\mathbf{minD}_C$, can give very fine-grained translations of $\mathbf{D}_P$ to the observed-score scale.

The validity argument should present both support for the logical consistency of the standard setting kernel with the types of test items included on the test and the accuracy of statistical procedures used to estimate the standard on the reporting score scale. It is tempting to use studies from the research literature on standard setting as support for the use of a standard setting kernel. When that is the case, it is important to show that the implementation of the standard setting kernel in the current standard setting process is similar to the process reported in the published articles to suggest that the results will transfer. Every standard setting kernel is implemented in a wide variety of ways. It is common to label the Angoff probability estimation kernel as a modified Angoff method, but all standard setting kernels are modified in some way when they are implemented. The use of many variations argues for detailed reports of the implementation of standard setting kernels and careful consideration of the similarity of implementation when using published reports and articles to support the use of a specific kernel.

9.2.2.2 Panelist Capabilities and Training

An important part of the validity argument is that the panelists who participate in the process can perform the tasks specified by the standard setting kernel to provide quality information that can be used to accurately estimate the standard intended by the agency. Most reports of a standard setting process will include information about how the panelists involved in the process were selected. Some, such as reports of NAEP standard setting (e.g., Loomis, Hanick & Yang, 2000), included the description of a nomination process with guidelines for selection of the types of persons who can be successful at the required task. The reports also included information about how individuals were selected from the pool of nominated candidates for specific standard setting activities. The goal from the perspective of the validity argument is to provide evidence that the panelists understand the information provided about $\mathbf{D}_P$, $\mathbf{D}_C$, and $\mathbf{minD}_C$, and can successfully perform the tasks required by the standard setting kernel after appropriate training.

A strong validity argument would make explicit the skills and knowledge needed by the panelists and provide evidence that those selected have those characteristics. That validity argument would also give details about the training process and evidence that the training was successful. Raymond and Reid (2001) give detailed guidance on both the selection of panelists and the training process. One approach to supporting the validity of a claim that panelists are qualified for the task is to include their credentials as part of the report on the standard setting process. Other supporting evidence is detailed information about the training process and information about how well panelists were able to perform on sample tasks related to the standard setting kernel. These sample tasks might be used as part of the training process.

Another type of evidence is acquired through evaluation surveys given to the panelists. Hambleton (1998) provides an example of an evaluation form. The evaluations usually cover (1) understanding of the performance descriptions, (2) the quality and success of training, (3) sufficiency of time, (4) understanding of the task, (5) factors used to make judgments, (6) appropriate level for performance standards, and (7) confidence in the results shown through feedback from the panelists.

Cizek (2012b) elaborated on the evaluation surveys suggested by Hambleton (1998). He agreed that appropriate and effective training, and panelists' feelings of success with the process, and confidence in the results are the important things to evaluate. He indicated that questionnaires given to panelists should be scheduled soon after the part of the process that is being evaluated. Cizek did not consider one evaluation at the end of the process sufficient support for the validity argument. In his chapter on evaluation of the standard setting process (Cizek, 2012b p. 171), he suggested six points in the process for evaluative questionnaires. These are essentially after every important activity.

9.2.2.3 Implementation and Documentation

Even the best of plans for a standard setting process can yield questionable results if the process is not implemented well. There are many threats to the validity of the interpretation of the results from a standard setting process. It may be that the definitions are carefully produced and precisely stated, the kernel that is selected has solid research support and is appropriate for the characteristics of the test used to operationally define the continuum implied by the definitions, and the panelists have the skills and knowledge to do the tasks required of the kernel and are well trained to do those tasks. But the amount of time allocated to the process may be too short, the facilitators may be untrained or not have the skills for their roles, and the administrative processes for implementation may be inadequate. Even worse, facilitators or others involved in the process may try to influence the results in directions that are different than those intended by the agency. Poor implementation can undercut quality plans for the standard setting process.

Hambleton and Pitoniak (2006) make recommendations for addressing implementation processes as part of the validity arguments. They list several criteria for providing procedural evidence for the validity argument. These criteria are (1) explicitness, (2) practicability, (3) implementation, (4) panelist evaluations, and (5) documentation. Explicitness means that the procedure is described in sufficient detail that it can be replicated. A detailed report of the process, including time schedules and roles for participants, allows a reviewer to determine if the process was implemented in a responsible way. Practicability means that the activities included in the plan have a reasonable chance of being implemented in a way that supports the entire process. For example, if the panelists are recording Angoff probability estimates on a form, are there sufficient staff and resources to input those values into data processing systems and perform the required analyses in the time available between rounds of the standard setting process?

Implementation means that the standard setting procedure was conducted in a way such that the goals of the procedure are clear to the panelists and all the component parts of the procedure are followed in a systematic way that leads to the thorough and accurate collection of data for the translation process. In other words, the persons managing the standard setting process did everything they could to properly conduct the process. Panelist evaluations refer to whether the panelists felt that they understood the requirements of the standard setting procedures and could implement them as intended. Documentation refers to how well all aspects of the standard setting process are described in reports of the results. Good documentation gives credibility to the entire process.

9.2.3 Internal Consistency

Pitoniak and Cizek (2016) interpreted the internal consistency part of the validity argument to mean that the results are replicable and consistent, meaning the results are similar over occasions or participants. Clauser, Margolis, and Clauser (2014) provide a useful summary of the types of analysis models used to investigate whether the results of a standard setting study are replicable and consistent. They classified analyses into three categories. The first category was the consistency of standards over replications of the process. Replicability of the process can be evaluated using parallel implementations of the same process or similar processes. The results from these implementations are frequently analyzed using generalizability theory to give estimates of the error in the estimated standards related to various components of the data collection design using the Brennan standard error statistic (Brennan, 2002) or generalizability theory estimates of standard error (Brennan & Lockwood, 1980).

The second category is intra-panelist consistency either across rounds of ratings or with other empirical data such as proportion correct estimates of item difficulty. Almost all standard setting processes that use multiple rounds of ratings by panelists report information about how the estimated standards change over rounds and the amount of variation within rounds. The third category is consistency between panelists. It has become standard practice to report distributions of panelists' estimated standards and measures of agreement among panelists.

Several researchers such as Linn, Madaus, and Pedulla (1982) and Cusimano (1996) have included the concept of decision consistency along with discussion of false positives and false negatives to the types of evidence for internal consistency. These concepts lead to consideration of loss functions (the costs of making incorrect decisions).

The research literature on investigations of the internal consistency of standard setting processes is extensive. To put some structure on the work that has been reported, it is organized here from fine-grained at the panelist level within rounds of ratings, to intra-rating consistency across rounds or with other empirical data, to full replication of the process. Then decision consistency results are considered.

9.2.3.1 Consistency between Panelists

Under ideal conditions, panelists have the skills necessary to perform the tasks specified by the standard setting kernel; there are clear $\mathbf{D_P}$ and $\mathbf{D_C}$ for the standard, and the reported test score is a good representation of the continuum implied by the policy definition $\mathbf{D_P}$. If that is the case, the translation of $\mathbf{D_P}$ to a point on the reporting score scale should yield a similar result for each panelist. Exactly the same estimate of the standard is not expected. Even the most competent translators make interpretations and word choices when translating from one language to another resulting in slightly different phrasing. The same is true with standard setting. But the variation in estimated standards should not be large. The literature on standard setting shows many ways for indicating the consistency of the standards estimated from the information provided by panelists. Unfortunately, there is little guidance about how much consistency is needed for the results of the standard setting process to be considered acceptable. More consistent is better, panelists who are outliers can be detected and may be removed from the process, and, if more than one standard is being estimated on the same reporting score scale, the distribution of estimated standards should not overlap and should certainly have statistically significant differences in location.

The most common indicator of consistency of estimated standards from information provided by panelists is the standard deviation of the estimated standards. Nedelsky (1954) used the standard deviation of estimated standards as a measure of "disagreement." He also compared the panelists' observed ratings of items to randomly generated ratings to determine if they were different. He concluded that there was evidence that panelists were not using a random process to generate their ratings. Cizek (1996), Hambleton (1998), and Plake and Cizek (2012) among others suggest the use of this same statistic, but they give it different labels and interpretations. Hambleton (1998) indicates that the standard deviation of estimated standards is a measure of consensus. Plake and Cizek (2012) use the standard deviation of estimated standards as "evidence of cohesiveness in the panelists' perceptions of the appropriate cut score for the test" (p. 185). One expert's measure of consistency is another person's measure of inconsistency. Shepard, Glaser, Linn, and Bohrnstedt (1993) used variation in panelists' judgments as a major reason for being critical of the Angoff probability estimation kernel used for standard setting on NAEP. Kane (1994) indicates that low spread of scores for the borderline group method is an indicator of quality. If the scores for the borderline group are tightly clustered, it supports the accuracy of selection of the group by the panelists.

Other standard setting experts have a more complex conception of consistency. Harrison (2015) used a statistic developed by van der Linden (1982) to assess intra-judge consistency. That index compares Angoff ratings to those predicted by the Rasch model with a correction for the difficulty of the item. The statistic is $C_{ij} = (\mathrm{Max}ES_{ij} - ES_{ij})/\mathrm{Max}ES_{ij}$, where i refers to the item, j to the judge (panelist), and ES_{ij} is the difference between the probability of correct response predicted by the model for individuals at the cut score and the rating given by the judge. $\mathrm{Max}ES_{ij}$ is the maximum difference possible between the predicted value and the value given by the judge. The index is on the probability scale. This statistic can be used at the item level across judges or for the full set of ratings. In this case, the measure of consistency includes the characteristics of the Rasch model as part of the definition of consistency.

A major issue with the interpretation of measures of panelist consistency is the criterion for level of consistency necessary to support the validity argument. The position taken in this book is that there is an intended standard implied by $\mathbf{D_P}$ and the goal of the standard setting process is to estimate that intended standard as closely as possible. Consistent with that position is considering panelists as translators of $\mathbf{D_P}$ to a point on the reporting score scale. There is also an expectation that qualified panelists will provide data that yields similar estimates on the reporting score scale because the panelists are translating the same $\mathbf{D_P}$. High consistency among panelists gives some support to the validity argument that the estimated point on the reporting score scale is an accurate translation of the intentions of the agency. When the results obtained from the information provided by panelists have high variation, this suggests that there are weaknesses in the validity argument. For example, Shepard, Glaser, Linn, and Bohrnstedt (1993) noted that there were large inconsistencies in the standards set by separate groups of panelists for NAEP. These inconsistencies led them to express lack of confidence in the results. They view it as important that panelists come to consensus on the standard (p. 128). The need for consensus is not supported by all persons planning and conducting standard setting. Some persons responsible for the standard setting process explicitly state that they do not encourage consensus (see Clauser, Margolis & Clauser 2014 and Schulz and Mitzel 2006 for examples).

Clauser, Mee, and Margolis (2013) explain why they do not encourage consensus. They believe that panelists may not interpret feedback about difficulty of test items in the same way examinees experience the difficulty of test items. Their view is that this can lead to convergence of consensus to the wrong values. Agreement among panelists may be high, but there may be systematic error variance. Clauser et al. further suggest that emphasizing consensus over connections to empirical data might lead to systematic error. They state that "caution is warranted when providing panelists with information about the judgements of other panelists" (p. 83). Such feedback "may systematically bias results" (p. 83).

Schulz and Mitzel (2006) had different reasons for not seeking consensus. They let each panelist define "minimally qualified" in their own way through the placement of the first bookmark in the Mapmark process. Then the panelists discussed the results but did not need to reach a consensus. Schulz and Mitzel based this approach on *The Wisdom of the Crowds* (Surowiecki, 2004) that proposed that collective thought is better than any one person if no one person is allowed to dominate the discussion. They also believed that independence of judgments will make mean cut scores more stable and that lack of independence means one panelist has influenced the others.

Although the different interpretations of the consistency of the estimated standards from panelists raise some interesting issues, from the perspective of a validity argument, having panelists reach some level of agreement on the location of the standard lends more support to the quality of the translation of $\mathbf{D_P}$ than wide variation in the estimates of the standard. Of course, the consistency of estimates is only one, and a relatively small, component of the validity argument. It is the support provided by the overall collection of evidence that is important.

9.2.3.2 Consistency over Rounds or with Other Data

The previous section discusses the consistency of the results obtained from the panelists. It is "within panel" consistency. The literature on standard setting also suggests evaluating the quality of the results from a standard setting study using "within process" consistency. This type of consistency is evaluated by observing how panelists' ratings change over the rounds of ratings and how the ratings compare to other information provided during the process, such as empirical estimates of the difficulty of test items.

The rationale for using the information about changes in the variation in estimated standards over rounds as evidence for the validity argument is that as panelists get a more thorough understanding of $\mathbf{D_P}$, $\mathbf{D_C}$, and $\mathbf{minD_C}$, and feedback on their initial efforts at translation, they will produce more precise translations and those translations will be more similar. Cizek (2012b) discusses the observed convergence of standards from different panelists over rounds of the standard setting process. If there is convergence toward a common standard, he considers this evidence that the standard setting process is working as intended. Yang (2000) makes the same point from the opposite perspective: "Low interrater consistency for a panelist indicates poor quality of ratings" (p. 4). She drew this conclusion based on analyses of consistency and reasonableness of ratings in the context of NAEP standard setting. She observed that the standards computed using data from some panelists had large changes from the second to third rounds of the process. Yang interpreted the changes as evidence that those panelists did not have a good understanding of their task. These results were also used to stress the importance of feedback so that panelists could gain a better understanding of the process and adjust their ratings, and the importance for monitoring the work of individual panelists for indications that they are having trouble with the process.

The second type of evidence from the standard setting process that is often labeled as evidence of consistency is the relationship between the information provided by panelists and empirical information about examinee performance. A common type of empirical information employed with standard setting methods that use ratings of individual items, such as the Angoff methods, is the proportion of correct responses to the items. Different standard setting studies use different variations of this type of information. Some use the proportion-correct information from the full sample of examinees who took the test. This is the traditional measure of item difficulty. Other studies use the conditional proportion (or probability) of correct response at the estimated standard. This value is computed directly either from the responses of examinees selected from a narrow range around the estimated standard or from an IRT model that is used to calibrate the items on the test. For any of these indicators of item difficulty, many standard setting researchers expect that a strong relationship between panelists' item ratings and the empirical measures of item difficulty is support for the quality of the standard setting process. Clauser, Mee, and Margolis (2013) explain the logic behind using consistency with external data as an evaluation of Angoff-type methods. "If judgments are logically consistent, there should be a close correspondence between the judgments and the empirical conditional probability of success on an item for an examinee whose score is at the cut score" (p. 72). They evaluated consistency by computing the correlations between panelists' ratings and the conditional probability at the estimated standard computed from the IRT model used to define the reporting score scale for the test. Clauser, Clauser, and Hambleton (2014) also emphasize that the relationship between empirical item difficulty and the rating of probability of correct response is an important part of the evaluation of a standard setting process.

But there is some controversy about how strong the relationship between item ratings and empirical measures of difficulty should be to support the validity argument. Demanding a high correlation as support for the validity argument implies that expected performance for difficult test items should be lower than the expected performance on easy items. But the information provided in $\mathbf{D}_P$ and $\mathbf{D}_C$ might not be consistent with the assumption that there should be a high correlation. For example, suppose that the agency responsible for certification of medical professionals determines that some difficult material is critical for successfully performing in the profession. Even though test items related to that material may be difficult, $\mathbf{D}_C$ may indicate that the qualified candidates show competence on that material. Proper translation of $\mathbf{D}_C$ using Angoff probability ratings should yield high probabilities for those items even though the empirical proportions correct may be low. To the extent that $\mathbf{D}_C$ includes skills and knowledge of that type, the correlation between ratings and empirical difficulty estimates will be reduced.

Clauser, Clauser, and Hambleton (2014) provide results that seem to show the conflict between the skills and knowledge described in $\mathbf{D}_C$ and the capabilities of candidates as shown by empirical estimates of performance on test items in different content areas. They reported a substantial range in correlations between observed ratings and those obtained from the estimated standard and the calibrations for the items from the IRT model (–.106 to .512). These correlations were from different content areas and panelists. Five of 29 panelists in the study had non-significant correlations between their ratings and the empirical estimates of difficulty. It is difficult to determine if the low correlations are due to poor estimation of item difficulty by panelists or good translation of the intentions of the agency for proficiency on difficult content.

Clauser, Kane, and Clauser (2020) further discuss this controversial issue. They argue that the variation in judges' ratings that are due to the empirical differences in difficulty are not error, but the deviations from what is expected from the empirical data should be considered as error. They discount the position that the panelists are indicating what the minimally competent examinee *should* do rather than what they *are* doing. Panelists' variance from the empirical results may be due to both error in judgment and different perceptions of the importance of different dimensions of the test.

Clauser, Kane, and Clauser (2020) use the squared correlation between empirical estimates of item difficulty and ratings to estimate the variance accounted for by the empirical differences in item difficulty. They report a correlation of .514 between the mean item ratings across panelists and the empirical difficulties. From that they estimate that 30% of the variance in ratings is due to the difference in difficulty of items and conclude that the other 70% of the variance is due to error. They do not consider that the magnitude of the correlation might be partly the result of the requirements in $\mathbf{D}_C$.

Wyse (2018) uses the relationship between Angoff probability ratings and empirical estimates of item difficulty in a slightly different way. He is concerned that panelists might be reluctant to use the extremes of the probability scale when providing their estimates of the probability that minimally competent examinees will answer test items correctly. He suspects that panelists may be regressing their estimates toward the middle of the probability scale. This work addresses an issue identified by McLaughlin (1993) in the context of NAEP.

Wyse (2018) used three indices to check for regression effects in the ratings from the use of the Angoff probability estimation method. These are the correlation between proportion correct values conditional on the estimated standard and the mean estimated probabilities from the panelists, the mean absolute deviations between those values, and the root mean square deviations. He had the same expectation as other researchers into standard setting that the correlation between panelists' ratings and empirical measures of difficulty should be high. He added an expectation that the deviation statistics should be near zero. He also suggested looking at the scatter plots of average item ratings with the conditional proportion correct values, and the average item ratings and the variance in ratings. The first plot shows the regression effects, and the second the amount of error in the estimates. His overall conclusion was that for the cases that he investigated, panelists were not adequately considering the differences between easy and difficult items. There were regression effects. These results indicated that panelists needed more training to sensitize them to the range of difficulties present in the items on the tests used in the standard setting study.

Overall, the literature on consistency over rounds of the standard setting process and with empirical item difficulty information indicate that convergence of estimated standards from panelists over rounds and at least a moderate relationship between ratings of items and the empirical item difficulty are evidence in support of the accuracy of the translation of $\mathbf{D_P}$ to a point on the reporting score scale for the test. No specific criteria are given in the standard setting literature on the amount of convergence or the strength of relationship with empirical difficulty that is needed to support the validity argument. However, there is likely agreement that no convergence over rounds (or divergence) and/or no relationship with empirical indicators of item difficulty would raise questions about the validity of inferences from the estimated standard.

9.2.3.3 Full Replication of the Standard Setting Process

Full replication of the standard setting process means that two or more distinct panels are convened, and the full process is conducted separately with the different panels. These types of studies have been done since the early work by Nedelsky (1954) who compared standards set on the same test a year apart and with different tests. He reported that the F/D grade cut scores were comparable under the conditions he studied. Hambleton and Pitoniak (2006) labeled these as parallel panels because the expectation was that the panels were equivalent samples from the population of panelists. They suggested using generalizability theory analyses with parallel-panel designs to get estimates of standard errors. In the simplest case of two panels, the Brennan (2002) method for estimating the standard error can be used. The expectation is that if the process is well designed and if parallel panels are recruited, the estimated standards should be similar.

Clauser, Margolis, and Clauser (2014) report on an elaborate study of the standard setting results from multiple panels. They found that there was a large variance component in some cases related to the selection of panelists. They discuss the amount of variation within panelists, between panelists within a study, and across studies. Studies that consider all three sources of variation are rare because of the expense of such studies. Many standard setting studies now use a split-panel design where the panel is divided into two or more subgroups that work independently but have common training and feedback. Then the standard setting results are compared across panels to given estimates of the amount of error in the process.

Boursicot, Roberts, and Pell (2006) used interclass correlations and variance in standards across groups as measures of what they labeled as "reliability" of the standard. The term "reliability" does not seem technically correct because the technical definition of reliability is the ratio of the true-score variance to the observed-score variance. If only a single standard is being set, the true-score (true-standard) variance is zero, so technically the reliability must be zero. Of course, if multiple standards are being set on the reporting score scale, there is variation in the true standards and the observed variance can be compared to an estimate of the true variance of the standards. This is similar to reliability estimated from analysis of variance (Hoyt, 1941).

Boursicot, Roberts, and Pell (2006) had the luxury of multiple panels, so the implied "true" score in their conceptualization of reliability was the grand mean over panels. This conception of the "true" standard is reasonable if all panels represent the population of panelists, and they are equally proficient at the translation process. If those assumptions are not met, the grand mean may not be a good estimate of the population value and it will not be possible to know which panel gave an estimated standard that was closest to the intentions of the agency.

The use of replicate panels, or at least splitting a panel into subgroups that function independently, is an effective way to determine if the result from a specific panel is idiosyncratic. Having replicate panels would offer a response to the concerns expressed by Shepard, Glaser, Linn, and Bohrnstedt (1993) that the estimated standards were too dependent on the characteristics of those selected to be on a single panel and thereby strengthen the validity argument.

9.2.3.4 Decision Consistency

The usual result from the application of a standard set on the reporting score scale for a test is the classification of examinees above or below the standard. The inference made from these classifications is that those individuals who are classified above the standard match the requirements given in $\mathbf{D}_P$ and $\mathbf{D}_C$, and those classified below the standard are lacking in some aspects described in the definitions. Decision quality refers to the degree that evidence supports that the classifications made using the standard have the desired meaning. Two indicators are frequently provided to give information about decision quality. One type of indicator provides information about the consistency of classifications made using the standard. That is, if the classifications were made using a different form of the test that operationalizes the continuum for the standard, or if the standard setting process were replicated with a different panel or process, would the same individuals be classified as above or below the standard. Decision consistency measures do not assess whether the standard is located at the correct point on the continuum. They only indicate the extent to which the classifications can be replicated when various aspects of the process have been repeated.

Decision accuracy refers to whether the correct individuals are classified above and below the standard. In the ideal case, the test is perfectly reliable for assessing the target population, it is a perfect operationalization of the intended continuum, and $\mathbf{D}_P$ is accurately translated to the reporting score scale. In the ideal case, all examinees are accurately classified above or below the standard. Of course, none of these conditions hold, so there are errors in classification of the examinees. Decision consistency measures indicate the impact of some of the sources of error—mainly errors in test scores and variation in placement of standards on the reporting score scale. Evidence for decision accuracy is provided by the full validity argument. Some of the evidence used to support the validity argument is information about decision consistency.

Cizek and Bunch (2007) indicate that Standard 2.15 from the 1999 *Standards* requires that decision consistency measures be reported when decisions about individuals are made using cut scores on a test. They describe several ways of computing decision consistency indexes in Chapter 16 of their book with an emphasis on methods that require data from a single standard setting panel or test administration. Information about the direction and amount of classification error could help the agency decide if they need to make an adjustment to the recommended cut score. The agency might make an adjustment based on a concern that the number of individuals inaccurately classified as passing or failing is too high.

Concerns about the rate of misclassification of examinees based on cut scores are not new. Shepard (1980) discussed some of the issues related to computing cut scores using contrasting groups and borderline groups. She summarized methods for estimating the cut score that minimize errors in classification, or a loss function related to the errors. Shepard investigated classical decision theory approaches along with Bayesian methods. Kane (1994) called these procedures "decision theoretic" (p. 428). Shepard (1980) criticized these methods because it is difficult to specify the parameters of the decision theory or Bayesian approaches. Work needs to be done to help standard setters make good judgments about parameter selection. She called for more research on utility functions and the weighting of classification errors. Jaeger (1989) found the information from decision consistency studies as more critical.

> The fundamental reliability issue in this context is the consistency with which students would be placed in the *same* categories were the decision to be based on their performances on a parallel form of the competency test, perhaps administered at a different time under similar but not identical conditions." (p. 487)

Brennan and Kane (1977) give the derivation of a measure of dependability of classifications relative to a standard. They based the statistic on generalizability theory assuming that persons are sampled from a large population and items are sampled from a large domain. Each person is assumed to have a universe score that is the equivalent of a true score and that score is compared to a mastery cut score. The dependability coefficient that they proposed is like a reliability statistic, but in place of variances of true and observed scores, they use expected squared deviations of the true and observed scores from the criterion score. They indicated that their statistic is different from one suggested by Livingston (1972) in that it considers variation in sampling from both persons and items. Livingston only considered the variation due to persons. They consider the statistic as the ratio of signal to signal plus noise and give a formal estimator for the statistic.

Brennan and Kane (1977) used the proportion correct metric for the test as the reporting score scale. They provided an example with a cut score of .7. The result was described as a consistent estimator of the standard for large samples. Their example yielded a dependability estimate of .817 for a 12-item test. They make a key point about their index that it includes an error term for the sampling of items. The dependability statistic depends on both the cut score and the number of items in the test.

A key point made by Brennan and Kane (1977) is that the dependability coefficient uses a squared error loss function relative to the cut score. The farther an examinee is from the cut score, the more serious the error of misclassification. They contrast this loss function with a threshold loss function where all misclassifications are considered equally in error. They do not support the use of the threshold loss function.

There appears to be agreement that decision consistency is an important aspect of the use of a standard. There is less agreement on how the evidence for decision consistency should be obtained. Decision consistency is included here as one of the many sources of evidence that can be used to support the validity argument for making accurate inferences from the applications of a standard.

9.2.4 External Consistency

While the internal consistency of the information provided by panelists is important, it provides limited direct support for the validity argument for the standard. Internal consistency indicates that panelists have a common understanding of the $\mathbf{D}_P$ and $\mathbf{D}_C$, that they tend to agree on the general characteristics of the translation of policy to a point on the reporting scores scale, and the test that operationalizes the score scale has sufficient reliability to yield replicable results. However, internal consistency does not directly indicate that examinees who are above the cut score know and can do the things described in $\mathbf{D}_C$. To support the desired inference that examinees who score above the cut score are accurately described by $\mathbf{D}_C$, it is useful to check if those classified as above the standard can do the tasks implied by the content definition of the standard. Cizek (2012b) gave examples of comparing students' classroom performance with the classifications made using the standard and comparing candidates' experience with their classification based on certification/licensure tests. Ideally, if a person is determined to be above the standard on the reporting score scale, that person should be able to do the things required by the $\mathbf{D}_P$. Unfortunately, collecting such supporting information does not seem to be a regular part of standard setting studies. Researchers into standard setting procedures have been asking for such information for many years.

Scriven (1978), for example, was critical of the lack of "backtracking" from the decisions made with a standard to observe whether examinees identified as competent are later judged as competent. The idea of checking on the accuracy of decisions was included in Standard 4.20 of the 1999 *Standards for Educational and Psychological Testing*, as reported by Cizek and Bunch (2007). The standard states that classification into categories with a substantive interpretation should be supported with sound empirical data. Shepard, Glaser, Linn, and Bohrnstedt (1993) reported that there was insufficient evidence for the validity of cut scores for NAEP (p. xxi). Their conclusion was based on a comparison of classifications of students into achievement levels by teachers and researchers and classifications based on the standards set on the NAEP assessment reporting score scales. Generally, the classifications by teachers and researchers were higher than those based on the NAEP results. They also compared student performance on NAEP to reported results for the SAT and Advanced Placement tests and concluded that the NAEP standards were set too high (p. 129). Shepard (1980) called for more research on how to resolve such mismatches between external information collected as evidence for validity and the results from the standard setting study.

Hartke (1993) also attempted to get evidence about the validity of standards set on NAEP from results implied by results from the SAT and the Advanced Placement tests. The evidence was related to the validity argument for a specific part of the $\mathbf{D}_P$ given by the National Assessment Governing Board, the agency that called for the standards on NAEP. The $\mathbf{D}_P$ for the "Advanced" standard included language that indicated that students above the Advanced standard exhibited "readiness for rigorous college courses." The evidence used by Hartke (1993) was difficult to interpret because NAEP and the two external tests have different administration contexts, different content specifications, and different levels of motivation. While there certainly should be some connection between the results from these assessment programs, the differences in the assessment programs make it difficult to determine how the evidence from the comparisons should be used.

Cizek (2012a) suggested that regression procedures could be used to get validity evidence about the classifications using standards that indicate college readiness. However, Kane (1994) indicated that such studies were seldom done because it is difficult to set a standard on the criterion that is being predicted and the sample used for the regression analysis does not have persons who do not meet the admissions requirements for colleges. These factors make it difficult to interpret the results of the regression analyses.

Kane (1994) discussed the issues related to using the results from different tests that are measuring similar things to the test that is the operationalization of the continuum implied by $\mathbf{D}_C$. He indicated that the results from similar tests may not support the validity argument if the alternative tests are less reliable than the test used in the standard setting study, or if those tests do not measure the same construct. Thus, even though consistency with information external to the standard setting study appears to be a useful source for validity evidence, it is difficult to gather compelling evidence. When results from the application of standards are compared to external information, the results are ambiguous because the other sources of information include construct-irrelevant variation or have different amounts of error variance.

Although the use of external evidence to evaluate the results of a standard setting process has intuitive appeal, as Kane (1994) indicates, there are many caveats when deciding the usefulness of such data. Some researchers, such as Shepard, Glaser, Linn, and Bohrnstedt (1993), emphasize the comparison to external information, and they question the accuracy of the estimated standard when the interpretations from external information do not seem to match those from the use of the standard. The best advice seems to be to use consistency with external information with caution and consider how that information fits into the overall pattern of evidence for the validity argument for the standard.

Another source of evidence for the validity of the location of the standard on the reporting score scale that is external to a particular standard setting study is the result from the application of a different standard setting kernel. Validity can be supported if different methods give comparable results, but there is an argument that the results will be different because of the distinct types of judgments. Kane (1994) indicated that comparable results support validity and differences raise questions. He also indicated that "it is generally difficult to evaluate the quality of different standard setting methods" (p. 450). Cusimano (1996) summarized research literature on comparisons of standard setting methods. The summary shows that standards from different methods can be quite different. One study on a certification examination showed that the Ebel and Nedelsky methods yielded a 19-percentage point difference in passing rate. But the different methods might not have been implemented with the same level of quality (Kane, 2001).

9.3 The Validity Argument

The validity argument for a standard has the task of supporting the inference or inferences that are made from the results of the standard setting process. In the simplest case, the inference is that persons who are classified above the cut score estimated from the standard setting process have the characteristics implied by the policy definition for the standard, $\mathbf{D}_P$. The support for that inference is obtained from the evidence for support for individual claims that make the logical connection between $\mathbf{D}_P$ and the inferences made about individual examinees. Table 9.1 provides a list of possible claims and the types of evidence that are needed to support each of the claims. This is not meant to be a comprehensive list. The list is an example of the types of claims and evidence that documentation should include to support the use of the results of a standard setting process.

The list of claims and supporting evidence in Table 9.1 may seem imposing. It seems that a large amount of effort is needed to collect all the information

TABLE 9.1

Components of a Validity Argument

Claim	Evidence
1. The policy definition for the standard, $\mathbf{D_P}$, clearly presents the characteristics of those who meet the standard	Careful reviews by the agency and other stakeholders indicate that $\mathbf{D_P}$ clearly and precisely indicates the intentions of the agency
2. The content-specific definitions for the standard, $\mathbf{D_C}$ and $\mathbf{minD_C}$, accurately reflect the requirements of $\mathbf{D_P}$	Careful review of $\mathbf{D_C}$ and $\mathbf{minD_C}$ by the agency to verify that the information is consistent with $\mathbf{D_P}$ and their intentions. Review of $\mathbf{D_C}$ and $\mathbf{minD_C}$ by content experts to determine if the language used will accurately communicate to those providing instruction/training and those constructing the test that operationally defines the continuum for the standard
3. Test specifications are consistent with $\mathbf{D_P}$ and $\mathbf{D_C}$	Content specialists and the agency review the test specifications to determine if they accurately communicate the intentions of the agency and precisely communicate the desired performance level to test developers
4. The test accurately reflects the requirements of the test specifications	An alignment study conducted to check to match between the test, or test forms, and the test specifications
5. The psychometric characteristics of the test support accurate classification of examinees in the reporting score range that will likely include the cut score	Perform psychometric analyses to determine the level of test reliability for the target examinee population and other psychometric characteristics that are consistent with the method for estimating examinees levels of proficiency
6. The selected standard setting kernel is consistent with the characteristics of the test, is practical given the resources for determining the cut score, and provides accurate estimates of the standard	The standard setting kernel is designed for use with the types of test items that are included in the test. Simulation studies show that if panelists correctly perform the tasks required of the kernel, the result is an unbiased estimate of the intended standard
7. The design of the standard setting ensures that panelists understand the task	A report of the standard setting process describes panelists training and feedback given to improve their performance of the required tasks
8. Panelists understand their tasks and competently perform them	Collecting evidence of panelists' understanding of tasks, feedback, etc. and statistical measures of dependability of results
9. Estimation procedures for the standards yield accurate results	Document the process for computing the estimates of the standards and present the rationale behind the estimation procedure
10. Examinees classified above the standard have characteristics that are consistent with $\mathbf{D_P}$	Information from sources independent of the test scores indicates that the examinees have the skills and knowledge implied by $\mathbf{D_P}$
11. The standard represents the intentions of the agency	The agency reviews all the information supporting the claims and determines if they support the validity argument that the standard represents their intention

needed to support the inferences desired from the use of a standard. However, much of the information listed in Table 9.1 is produced over time during the process of designing and implementing the standard setting process. The major effort required is collecting the information and organizing it into a report of the standard setting process. For example, the standard setting process typically has an agenda and materials used to orient panelists and train them for their tasks. This is the material required for #7 in the list. Likewise, there must be a method for computing the cut score based on the information collected from the panelists. That is #9 on the list. It should not be a difficult or time-consuming task to provide a report of the estimation process. With the foreknowledge that evidence is needed to support the validity argument for the inferences made from the adopted standard, the collection of the evidence becomes a normal part of the standard setting process.

Some of the claims listed in Table 9.1 are aspirational. Many standard setting studies do not have a clearly stated $\mathbf{D}_P$ as a starting point. That definition is elaborated throughout the process. Few standard setting studies include evidence that examinees who exceed the standard have all the skills and knowledge implied by the $\mathbf{D}_P$. That evidence cannot be collected until after the standard is set, and it does require extra effort to track passing candidates and get information about their performance. Professional organizations may get such information in an informal way through complaints, or lack thereof, that new hires do not have the necessary skills for the job. Such informal feedback may result in a future increase in the level of the standards.

It is now common for the agency that calls for the standard to receive a report of a standard setting process as part of the information they use to decide whether to adopt the standard. Organizing that report in a way that supports the validity argument will help the agency make its decision regarding adoption of the standard, and it will also provide a ready set of materials to support their decision.

10

Trends in Standard Setting Design and Development

10.1 Introduction

Like many other areas of education and psychology, the procedures used for translating policies into locations on test score scales are not static. This book contains descriptions of numerous standard setting procedures starting with the proposal from Nedelsky (1954) to variations suggested in 2022 when this book was being completed. No doubt there were procedures used prior to 1954 that did not get into the archival literature or that were classified under different keywords such as "grading" or "marking" practices. Standards have been set on tests going back to the Chinese civil service examinations as early as the year 622 (Teng, 1943) and in other places such as the British Royal Navy in 1677 (Cavell, 2010), but these standards were typically in the minds of the persons evaluating performance on the test rather than as points on the reporting score scale of the test. Each test form was different, and the expectation was that those persons doing the evaluations (e.g., flag-officers for the British Navy) would adjust for the differences. It was not until the development of large-scale assessments in the early 1900s (i.e., the Army Alpha and Beta Tests) and the psychometric theory to support the development of equivalent forms and equating procedures that the technical aspects of standard setting began to develop. Prior to that, standard setting was an administrative procedure rather than a specialized methodology.

Since 1954, many different standard setting approaches have been suggested (called *kernels* in this book) and there are numerous variations for each of them in reports of standard setting processes. One of the goals for this book was to present a framework for evaluating these options so that informed decisions can be made when selecting from the multitude of standard setting procedures and variations. A hope for this book is that the framework is sufficiently general that it can apply to procedures that will be proposed in the coming years. It seems foolhardy to try and predict what those new procedures will be, but there are current trends that are likely to continue to evolve for some time. This chapter describes four of those trends:

DOI: 10.1201/9780429156410-10

(1) refining the policy and content definitions for standards; (2) placing more than one standard on a reporting score scale or on linked score scales at different levels; (3) extending methodology for use with more complex test tasks such as simulations and technology-enhanced assessments; and (4) connecting policy and content definitions to the design of test tasks and test forms to enhance setting of standards.

Each of these trend areas will be described in turn, but that is not to suggest that they are independent. All of them are interconnected and the connections are important. As policy and content definitions are refined, they more clearly indicate the kinds of information needed from tests to support accurate decisions. The refined definitions make the connections to test specifications more obvious. The use of new, computerized item formats also needs to be connected to the definitions and test designs. Further, when multiple standards are placed on the reporting score scale for a test, there are clear implications for the level of accuracy needed at points along the reporting score scale to support accurate classifications into categories defined by those standards. The overall implications of these trends are that assessment policy, standard setting, and test design/development need to be carefully coordinated.

10.2 Policy and Content Definitions

The position taken in this book is that standards are set by the individuals who make policy, the agency, and their intentions are given in the policy definitions $\mathbf{D}_P$ for the standards. All other aspects of a standard setting process are mechanisms for translating the intentions as described in $\mathbf{D}_P$ into a point on the reporting score scale for a test that was judged or developed to be consistent with the intentions of the policy. An implication of this view of standard setting is that the precision and clarity of $\mathbf{D}_P$, and the other definitions that are derived from it (there are now many types: range descriptions; achievement-level descriptions [ALDs]; performance level descriptions, PLDs; borderline performance descriptions; etc.) are critical to obtaining an accurate translation onto the reporting score scale.

Perhaps in acknowledgment of the importance of these written definitions of the standard, those involved in developing standard setting procedures have begun the challenging work of producing guidelines for the text of the definitions and examples of definitions that they believe will lead to quality standard setting results (e.g., Egan, Schneider & Ferrara 2012; Hambleton & Pitoniak 2006; Schneider, Huff, Egan, Gaines & Ferrara 2013; Wolkowitz Impara, Buckendahl 2017). Developing policy and other definitions that facilitate the translation of the standard to the reporting score scale is challenging because $\mathbf{D}_P$ are written to communicate with a variety of audiences, and it seems desirable to keep them brief. Often, such as the case for NAEP standards, $\mathbf{D_P}$s are purposely kept general so that they can apply to all curriculum areas typically taught in elementary and secondary schools. For the NAEP case, the $\mathbf{D_P}$s are translated into content-specific achievement-level descriptions, ALDs, for different subject matter areas such as Mathematics, English Language Arts, Science, etc. The same challenges are encountered in some certification/licensure programs that have several specialty areas. The agency provides general $\mathbf{D_P}$ defining meaning of competency and then that general definition is applied to all the specialty areas.

While the work on improving the quality and precision of the language of definitions is likely to improve the results from standard setting processes, it is unclear if the recommendations from researchers/developers will have an impact on the policy definitions produced by agencies. Those who implement standard setting processes are typically hired by the agencies and improving the $\mathbf{D_P}$s requires influencing those who are in charge of the entire process. Persons who are in positions of authority over the process need to be informed of the importance of the language of the policy definitions and how the definitions drive the entire standard setting process. It will be interesting to observe whether the work on improving $\mathbf{D_C}$ and $\mathbf{minD_C}$ will have any effect on improving $\mathbf{D_P}$.

10.3 Setting Multiple Standards

Early applications of standard setting procedures tended to be in support of dichotomous decisions. Conaway (1979) described the use of standards to make mastery/non-mastery decisions in the context of educational programs. Similarly, many certification/licensure programs use standards to make competent/not competent to practice decisions. Even though Nedelsky (1954) developed his method in the context of assigning grades to students in a course, he concentrated on the cut score between grades F and D. He indicated that he had "few data bearing on an extension" to more grade points. He wrote that modifications to his method would be necessary for it to be applied to more grade levels.

10.3.1 Multiple Standards on the Same Test

Since the early applications of standard setting methodology, there has been a clear policy trend toward setting multiple standards on the reporting score scale for testing programs. The National Assessment Governing Board (NAGB) may have stimulated this trend by setting policy to report NAEP results using three achievement levels—Basic, Proficient and Advanced. Bourque (2009) reported that NAGB developed the $\mathbf{D_P}$s for the three levels in 1990. School-based testing was not the only place where multiple levels of standards were desired. Language testing to document the competency of persons on languages other than their native languages has also used many classification levels for reporting results. Several tests use the definitions provided by the Common European Framework of Reference (CEFR) as a basis for reporting results. The CEFR defines six levels of language competence (Council of Europe, 2001).

Persons implementing standard setting procedures quickly noted the challenges that arise when setting multiple standards on the same test. Plake (1998) points out that when the $\mathbf{D}_P$ calls for multiple cut scores, such as with NAEP, that has important implications for the design of the standard setting process. "It is possible, but difficult, to imagine a single panelist whose task it is to sequentially envision candidates whose level of performance is barely acceptable, not just for one point on the ability scale but for three different decision points" (p. 78).

Plake and Cizek (2012) suggest a solution for the burden placed on panelists by designing a modification to the Angoff item-score kernel called the "yes/no method." They indicated that the modification supported defining "hierarchical levels of performance" (p. 89). The modification to the Angoff item-score kernel requires panelists to determine the lowest cut-score where examinees will likely answer an item correctly (score of 1) and then impute that all examinees at higher levels will also answer the item correctly. The item scores for any levels below where panelists put their first "yes" are imputed as incorrect responses (score of 0). This reduces the number of judgments that panelists must make. "This feature of the Y/N method reduces the burden on the panelists for the number of independent decisions that need to be made for each of the items that comprise the test" (p. 189). Plake and Cizek describe a way of producing a matrix of 1's and 0's that shows the imputation of 1's and 0's. The cut scores on the number-correct score scale are computed by summing columns of the matrix. The matrix has items as rows and cut scores as columns. Plake and Cizek include several rounds of ratings with feedback such as *p*-values and impact data between rounds. They argue that for the setting of multiple standards, the Y/N method takes less time than alternative methods such as using the Angoff probability-estimation kernel to obtain probability estimates for an item for examinees at each of the standards. Data entry is expected to take less time as well.

Other standard setting kernels such as the Bookmark and Angoff probability-estimation kernels have been used for setting multiple standards on the same test. Some newer methods, such as the "generalized examinee-centered method" (Cohen, Kane & Crooks, 1999), have also been developed. Although there has been some research on determining the amount of error in the estimated cut scores, there is little detailed research on how well the different approaches to setting multiple cut scores achieve the goal of recovering the standards implied by the policy definition $\mathbf{D}_P$. One issue of concern is that the data used to estimate the multiple standards do not meet the statistical characteristic of "independence." Plake and Cizek (2012) saw this as a positive feature of the modified yes/no method because it reduced the burden on the panelists. However, they did not consider the price paid to reduce the burden.

A simple example is provided here to show the possible effect of dependencies in the information provided by panelists. This example is not meant to provide conclusive results because it is based on strong assumptions about how panelists respond to the process for setting multiple cut scores and the results are based on a single simulation without any replications. The intent is to stimulate interest in this issue so that more comprehensive research will be done based on either actual standard setting process data or simulations with design features that are supported by observations from panelists doing the standard setting tasks.

The example is for setting four standards on a 30-item test. The results are based on a three-parameter logistic model calibration of the test items. The item parameters are assumed to be well estimated and are treated as true parameters. Table 10.1 shows the standards intended by the agency that calls for the standards on both the θ-scale and on the true-score scale for the test. The values in the simulation are based on the standards set on a state assessment that reports results in five reporting categories here labeled simple 1, 2, 3, 4, and 5—5 being the highest.

The rows labeled "Angoff Probability with Error" and "Yes/No with Error" show the results for a simulated panelist if each cut score is treated as an independent standard setting. Note that the Angoff probability-estimation kernel gives very good results with only minor differences from the intended standards. The Angoff item-score kernel (Yes/No) shows the estimated cut scores tend to be more extreme than the cut scores intended by the agency. This feature of the Yes/No method was discussed in Chapter 5. The last two rows of the table provide an example of one way dependency might enter into a panelist's approach to providing information for estimating multiple standards.

TABLE 10.1

Example of Dependency Effects When Setting Four Cut Scores on a Test

	Cut Score			
Condition	**1–2**	**2–3**	**3–4**	**4–5**
Intended θ	–.75	.15	.75	1.25
Intended true score	15.8	19.3	21.7	23.8
Angoff probability with error	15.8	19.1	21.4	23.8
Yes/No with error	14	17	21	25
Angoff probability with error and dependency	15.8	20.0	22.5	24.6
Yes/No with error and dependency	14	18	23	27

To show the possible effects of dependency between ratings, the simulated panelist first estimates the probability of correct response for the items in the test for the first cut score with the rating error model based on the beta distribution that was used in Chapter 5. Then, if the estimated probability of correct response was greater than .5, it was recorded as a 1 for the Yes/No method and 0 otherwise. For the first cut score, this gives the identical results as the independent standard setting. The probabilities were then estimated for the second cut score, but if because of error the probability for an item was less than that for the first cut score, the probability was changed to be equal to the probability set for the first cut score plus the average difference in probabilities for the estimates for the two cut scores. After the adjustment, the probabilities were converted to 1 or 0 for the Yes/No method. The same process was followed for the third and fourth cut scores.

The logic behind this simulation is that after a panelist estimates the probability or the item score for the first cut score, the value for the next cut score must necessarily be higher because it represents greater proficiency. However, the panelist might overestimate the first probabilities and/or underestimate the second probabilities resulting in an unacceptable ordering of probabilities. When a panelist notices this unacceptable order, they will adjust ratings, but because the work for the first cut score has already been completed, the adjustment is made on the probability estimate for the second, and subsequent, cut scores. The result is that the estimation errors tend to accumulate in a positive direction resulting in higher estimated cut scores beyond the first one. Of course, the opposite effect could occur if the panelist started the process with the highest cut score and worked down to the lower cut scores. These possible effects are due to the dependencies in the estimates provided by the panelists.

One way of dealing with this issue is to have panelists start the probability-estimation process with the middle standard and then work out to the upper and lower cut scores. This approach would reduce the accumulation of error. However, no formal research studies were found in the standard setting literature that investigated the consequences of dependencies on standards when multiple standards are set. Given the number of occurrences of tests with multiple cut scores, this is a critical area for research.

10.3.2 Coordinated Standards for Tests at Different Levels—Vertical Articulation

A second trend for setting multiple cut scores appears in the literature on standard setting in the context of testing within an educational system. Tests used to evaluate student learning often have the same, or similar, content at different grades levels. For example, mathematics tests may be developed to assess achievement and growth in achievement in mathematics for grades 3 through 8. Standards may be set on each of the grade-level tests and the results compared to see if students in the grades are performing at levels that will lead to meeting instructional goals for the end of elementary education. Shepard, Glaser, Linn, and Bohrnstedt (1993) argued that the pattern of proportions above standards by grade should have what they labeled as "grade to grade coherence" (p. 100). They also suggested that the logical connections between standards at different school grades should be considered as part of the standard setting process (p. 125). However, they did not indicate how to determine the appropriate pattern across grades. They knew what they did not believe was appropriate—one grade level with standards much higher or lower than another. They suggest a simple approach of selecting percentile ranks of 25, 75, and 95 for three proficiency-level standards and using the corresponding scores on the reporting score scale as the standards for each of the grades. Of course, this simple approach assumes that percentile ranks have the same meaning at each grade level and that students in the same percentile range over grades have equivalent levels of competence over the instructional content for the grade.

More recently, other terms have been used to refer to the relationships between standards on different grade-level tests and the approaches to achieve "coherence." Lissitz and Huynh (2002) used the phrase "vertically moderated standard setting" to describe the process of setting standards at several grades that met the criteria for "coherence." Cizek and Bunch (2007) also use that term. Ferrara, Johnson, and Chen (2005) used the term "vertical articulation" to describe the process for developing appropriate connections between the standards set at different grades.

If "coherence" or "vertical articulation" is considered desirable for the standards set on tests of the same subject matter area at different grades, then an important question is how to estimate the standards on the reporting score scale so that they are accurate translations of the $\mathbf{D}_\mathbf{P}$'s and meet the requirements to be judged coherent. When coherence was first raised as an issue—Cizek and Bunch (2007) indicated that vertical articulation was a response to Federal legislation, mainly No Child Left Behind—the initial solution was to "smooth" the standards over grades to reduce inconsistencies. That is, standards were set independently for each grade level and then a cross-grade panel was convened to adjust the standards to approximate "coherence." Pitoniak and Cizek (2016) described the process in this way: "Once initial cut scores are recommended at the individual grade levels, a process called *vertical articulation* (or *vertical moderation*) is employed to smooth out inordinate or unreasonable fluctuations in the system of cut scores across the grades" (p. 42). Cizek and Bunch (2007) indicated that these fluctuations may be due to differences in the standard setting procedures used at different grade levels, differences in levels of content descriptions for the levels, and/or the regular sources of error variance in any standard setting study. They also relate this to vertical equating, vertical scaling, or vertical linking (p. 251). Thus, in their view, even if the panelists have good intentions and the tests are properly scaled, there are likely to be inconsistencies in where the standards are set, and these inconsistencies result in lack of coherence.

The comments on vertical articulation highlight the many issues that are present when an attempt is made to have standards coordinated over grades. Usually, standards on all the grades are set at the same time. That means that the sample of students taking the tests are cross-sectional rather than longitudinal. The characteristics of the samples of students may be changing as they progress through the educational system. The contents of the tests also change over grades to match the emphasis of the instruction and instructional goals. The reporting score scale may be defined for each grade level, or an attempt may be made to have a common scale over all the grades to facilitate the tracking of growth. A common scale is usually developed using a psychometric methodology called "vertical scaling." These are additional technicalities beyond that of producing definitions of performance that form a reasonable progression over grades.

Cizek and Bunch (2007) called their approach to this issue vertically moderated standard setting (VMSS). This methodology is supposed to derive a "set of cross-grade standards that realistically tracks student growth over time and provides a reasonable expectation of growth from one grade to the next" (p. 253). Cizek and Bunch (2007) indicated that VMSS is based on two assumptions: (1) there is a generic $\mathbf{D}_P$ that applies to all grade levels; and (2) the change in performance for the minimal qualified person should follow a regular pattern with instruction. The actual method for VMSS is judgmental smoothing of the relationship between the percent above a standard and the grade level. External information is used to inform judgments about which percentages are out of line with the others. The type of smoothing that is applied depends upon the relationship that is believed to hold for the connection between grade level and the standards. The assumed relationship might be a linear one with a constant increase in expectations from grade to grade. Alternatively, the assumed relationship might be nonlinear with increasing or decreasing differences in standards from grade to grade. There may also be expectations about how low-performing students are assisted to improve their performance, or how high-performing students are provided with advanced material. If these assumptions and expectations are correct, the score distributions will change in important ways as students progress through the grades. For example, the bottom of the distribution may be truncated because of successful remediation of poor performing students, and the top of the distribution may be skewed because of enrichment programs for high-performing students. Such shifts in distributions of performance will have an impact on the smoothing process.

Cizek and Bunch (2007) acknowledge the complexities caused by these assumptions and expectations. To address these issues, they suggest that "any application of VMSS is hampered if it is not supported by a theoretically or empirically sound model of achievement growth" (p. 273). It is not clear if this model should be developed by the panelists who are smoothing the standards or if the model is provided as part of the $\mathbf{D}_P$ produced by the agency calling for the standards. If the panelists are expected to develop the model, it is likely that they will try to find one that best fits the information they are provided. If the standards set at the various grades suggest a curve, the panelists will likely emphasize the curve through smoothing. If the agency is expected to provide the model as part of the $\mathbf{D_P}$ for the standards, the members of the agency will need to be knowledgeable about reasonable models for learning over grades.

As with all standard setting processes, there are many variations in the way that VMSS is implemented. VMSS may be part of the standard setting process at each grade level with feedback to panelists about where standards are set by other grade levels. That feedback is used to moderate the standards at a particular grade level. There may also be use of the standard error estimates of standards to guide the magnitude of adjustments that are needed to smooth the results. VMSS may also be completed through a separate process with external data used to inform decisions. Formal statistical data fitting may be used to recommend the model that is most consistent with the percent above the standard over grades.

Cizek and Bunch (2007) highlight a potential problem with the smoothing process. The adjusted standards may not match the content $\mathbf{D_C}$s for the standards. "From both theory-development and practical perspectives, the issue of how to maintain the meaning of cut scores and fidelity to PLDs is probably one of the most fundamental for future research to address" (p. 273). Standard setting on its own is a challenging process. VMSS adds substantially to the level of challenge in the process.

To avoid the issue of mismatch between content definitions and smoothed cut score locations, Lewis and Haug (2005) included alignment across grades as part of the $\mathbf{D_P}$. Their approach to vertical articulation required that the agency include expectations about the pattern of standards over grades as part of the $\mathbf{D_P}$. Then the content definition, $\mathbf{D_C}$, and $\mathbf{minD_C}$ are developed to be consistent with the expected pattern of performance. These definitions need to be developed prior to the standard setting process so that panelists can use them when performing the tasks required of the standard setting kernel.

Although the approach suggested by Lewis and Haug (2005) addresses the problem of match between content definitions and the pattern of standards over grades by working out the expected pattern before the content definitions are developed, determining a reasonable pattern before the standard setting process is implemented remains a challenge. Lewis (2002) provides perhaps obvious initial guidance. He believes that if there is a vertical scale across grades, "that the proficient cutscore should increase with grade" (p. 3). To further assist the agency in including the pattern of standards into the $\mathbf{D_P}$, Lewis and Haug (2005) convened a meeting of educators in the subject matter area to develop a recommendation for the model of increase in standards across grades. The educators reviewed all previous data to help develop the model. As part of that process, Lewis and Haug (2005) raised the issue of the agreement between the $\mathbf{D_P}$ and the $\mathbf{D_C}$s for the standard. The recommended pattern (model) for the standards was then submitted to the agency for possible inclusion in the $\mathbf{D_P}$ for the set of standards. The requirement for the agency to include the pattern of progression of standards in the $\mathbf{D_P}$ highlights the need for the agency to be well informed about the broader context for the standards. Lewis and Haug (2005) added an additional step to the standard setting process to help inform the agency about important aspects of the standards.

In the previous section of this chapter, the challenges encountered when setting multiple standards on the same test were described. With vertical articulation, the challenges are even greater. Lewis (2002) and Lewis and Haug (2005) have developed designs for the standard setting process that have the goal of integrating vertical articulation with the use of a Bookmark kernel. Lewis (2002) described a standard setting process where panels provided information for estimating standards for multiple grades using ordered item booklets containing items from all those grades placed on the same scale. The assumption was that the construct being assessed was the same over the grades that were combined. This allowed standards to be set on the tests for the multiple grades in the same standard setting session. The approach allowed evaluation of the pattern of standards during the standard setting process. Panelists could see the progression of standards and determine the match to the pattern included in the $\mathbf{D_P}$.

Lewis and Haug (2005) make the same assumption about a common $\mathbf{D_P}$ over grades that was made by Cizek and Bunch (2007). Lewis and Haug (2005) state: "The proficient cut score set at one grade of a multi-grade assessment program begs to be consistent with the message sent by the proficient cut score at other grades within the same content area" (p. 12). Further, "we assume that the role of the cut scores within the accountability model and the consequences associated with students placing in the various performance levels are similar across grades" (p. 14).

One requirement of this approach is that panels provide ratings on items from more than one grade: "participants work together in groups that simultaneously consider cutscores for several grades and make judgments such that the relationship of the cutscores at two or more grades is explicit" (p. 4). This means that panelists need familiarity with the students and curriculum for more than one grade. In the example provided by Lewis (2002), all panelists had direct experience with the range of grades that were included in the ordered item booklet for a bookmark procedure.

When implementing a standard setting process that involved vertical articulation, Lewis and Haug (2005) made several important design decisions. They were (1) suggesting several alignment patterns for standards across grades and allowing panelists to recommend one of them, (2) providing predicted locations of standards based on the selected alignment pattern as a starting point for panelist ratings, (3) providing feedback about the standards set at other grades, and (4) encouraging discussions about coherence of standards during the standard setting process. The process was challenging and there was a need for negotiation of standards at some grade levels. Some panelists were not satisfied with the results when they tried to match the requirements of policy, content, and pattern of standards over grades.

The initial results from Lewis and Haug (2005) clearly indicate the need for much more research into the process for producing vertically articulated standards. As highlighted by the researchers in this area, there are many potential sources of variation when attempting to produce vertically articulated standards. The process depends on the quality of the vertical scaling of the tests, the policy definition for the pattern of change in the standards over grades, the match of the pattern to the content definitions for each grade, and the capabilities of panelists to understand their tasks. These topics, and more, suggest a research agenda that can continue for years.

10.4 New Approaches to Standard Setting

Persons working in the area of standard setting are continually trying to improve on the process so that convincing evidence can be collected to support the inferences that users of a standard want to make about the capabilities of examinees. Most often, the improvements are largely unnoticed because they are hidden in the details of reports of a particular standard setting process. There are endless modifications to existing types of standard setting kernels. Less frequently, developers suggest more dramatic changes to the process of setting standards. These efforts may result in a new standard setting kernel with a new label. There is no organization like those in biology that decides when a new species of plant or animal has been identified that determines when a new standard setting process is uniquely different than those that already exist, or if the suggested name is appropriate. A standard setting process is considered as new when other researchers in the field decide to reference the process as something new. Two such processes are described here that were "new" when this book was completed. Time will determine if they are added to the list of operational standard setting processes.

10.4.1 Embedded Standard Setting (ESS)

Lewis and Cook (2020) proposed a standard setting method that has some similarities to the Bookmark kernel, but that moves the translation process directly to the development of test items. The process does not convene panels to get information to translate content definitions to the reporting score scale. Rather, the method transfers the translation process to the persons who develop the test items for the test. The goal of the process is to make a strong connection between test design and development, and the estimation of the standard on the construct defined by policy. To accomplish the goal, strong assumptions are made about the quality of $\mathbf{D}_C$'s, which Lewis and Cook label as achievement-level descriptions (ALDs), and the capability of item writers to produce items that specifically target the skills and knowledge implied by the ALDs. The method was specifically designed for the situation when there are multiple standards on the construct implied by the $\mathbf{D}_C$. Those standards divide the scale for the construct into regions and each region has its own $\mathbf{D}_C$ that Lewis and Cook labeled as a "range ALD." Then item writers produce items to measure the specific content described in the range ALDs. The ESS method then assumes that the items written to the ALDs form distinct ordered difficulty sets.

To determine a cut score using the ESS method, the items written to match the requirements of the range ALDs are calibrated using an IRT model. Then the items are ordered according to a mapping criterion in the same way as is done to form an ordered item booklet for the Bookmark method. Lewis and Cook (2020) used the traditional mapping criterion of .67, but they note that the selection of a different mapping criterion will yield different estimates of standards. The mapping criterion is the definition of what it means to "know" the information required of an item written to match the requirements of a range ALD. Rather than specify a mapping criterion, Lewis and Cook suggested that it be determined from the IRT scaling as the value that minimizes inconsistencies.

For the ESS method, item writers need to produce test items for at least two levels—above and below the cut score. The cut score is determined by finding the location on the IRT scale that minimizes the misclassification of items (the inconsistencies) into categories above and below the cut score based on the target range ALDs item writers used as a guide to producing the items. The cut score is the mapped point on the scale for the item that minimizes the inconsistencies. If more than one point exists that minimizes the inconsistencies, the cut score is the point that is more consistent with the content of the ALDs. Lewis and Cook suggested that a small committee of subject matter experts be convened to review the inconsistent items and resolve the location of the standard and the inconsistencies. Their review would also include consequences information. Real data analyses showed that as many as 25–30% of items may be inconsistent. In best cases, the percent of inconsistent items is 10% or less.

Lewis and Cook (2020) used the correlation between the item locations and the classification level as a measure of the quality of the standard setting process. Their simulated examples had correlations in the high .80s or the .90s. Real data examples based on 14 language tests with multiple cut scores had correlations that range from .20 to .87 with most in the range from .70 to .87. They also discussed using impact data as a means of determining the reasonableness of the estimated standards. When empirical data about the performance of test items does not match the expectations of the item writers, when the mapped items do not fall into distinct ranges according to the ALDs, they suggested three explanations. One was that the item writers incorrectly targeted the items. A second was that the item brings in construct-irrelevant material that influences the difficulty. The third was that the ALDs do not reflect the pattern of what students can do. Adjustments to the process were suggested for each of the cases that were believed to hold.

Lewis and Cook (2020) made an argument that if there are many inconsistencies in the classification of items according to the achievement levels versus empirical data, then the test development process was flawed, and other standard setting kernels cannot be meaningfully applied either. The consistency of classifications is part of a validity argument for the estimated standards.

> Given the premise that items representing achievement levels should be clustered together based on empirical difficulty, then the validity of the established cut scores is dependent on the extent to which the KSAs [knowledge, skills, and abilities] conferred upon each achievement level by the item mapping align with the claims asserted by the ALDs for that level. (p. 18)

They also made an argument that with a large pool of items, the items should be selected so that they match the classification by ALDs and that those that do not match should not be used.

The ESS kernel is unique among commonly used standard setting kernels in that it does not require convening a panel of experts to provide information for estimating the standard. The ESS kernel relies on the quality of test items as accurate translations of specific components of the $\mathbf{D}_C$s. In a sense, this kernel is the equivalent of word-by-word translation from one language test to another. The assumption is that if every word is accurately translated, the meaning of the full text will be accurately conveyed as well. The check for inconsistencies and resolving those is doing some editing (smoothing) of the final translation. The key to the use of this kernel is how well the test items written for a range ALD represent the skills and knowledge implied by the range ALD. If there is a high level of confidence in the item development process, this standard setting kernel moves the standard setting process into the test development process and eliminates the need to have a separate standard setting study. Research studies are needed to show when the item development process is adequate to support the accurate estimation of points on the reporting score scale.

10.4.2 Item Descriptor (ID) Matching Method

Ferrara, Perie, and Johnson (2002) started with the same premise as the ESS method, but they did not have the same level of confidence in the item development process as Lewis and Cook (2020). Their method, called the Item Descriptor (ID) Matching Method, continues to use a panel of experts to make the connection between test items and the skills and knowledge needed to be considered above a standard. The method involves matching test items from a test to the content descriptions for each level above a standard. They use the term "performance-level description" (PLD) for the content descriptions rather than ALD. As with the ESS method, detailed PLDs are extremely important. These descriptions need to be detailed and the agency should approve them before use. The method as implemented by Ferrara, Perie, and Johnson (2002) used the same kind of ordered item booklet (OIB) that is used for the Bookmark kernel. Starting with the easiest item, panelists are asked to indicate which PLD best matches the content of the item and the skills needed to answer it. This is done for each successive item. This can also be done for each score point for polytomous items.

Although the items in the OIB are ordered in difficulty according to a mapping criterion, the panelists do not have to use that ordering when classifying items as matching PLDs. There may be regions of the OIB where classifications are mixed between lower level and higher level PLDs. These are called threshold regions. Mixed classifications (inconsistencies) can represent uncertainty in the ordering of the items in the OIB or in panelists' classifications or both. Threshold regions are defined as

> beginning [at] the first item after a run of at least three consistent classifications at a lower performance level and ending with the item just before a run of at least three consistent classifications at the next (i.e., higher) performance level. (p. 10).

Cizek and Bunch (2007) remarked that this rule is "somewhat arbitrary" (p. 204) and that research should be done to check the rule. Also, the rule does not work well with short tests.

There is no single recommended way of setting a cut score using this method. Ferrara, Perie, and Johnson (2002) emphasized having panelists select a point in the threshold region as the estimated cut score. Alternatively, the midpoint of the locations of items in the threshold region could be used, or logistic regression analysis can be used to statistically estimate the point where items are equally likely to be classified into the higher or lower PLD. Cizek and Bunch (2007, p. 201) suggested using the mean or median of the scale values for the items in a threshold region.

After a first round of identifying cut scores, panelists discuss the differences in classification of items with PLDs. Then the panelists can reconsider their classifications and the location of the cut scores in possibly new threshold regions. There may be a third round with impact data after the second round.

As with any of the standard setting kernels, the ID Matching kernel can be implemented in many different ways. There has not been sufficient experience with the method to determine the combination of specific details such as mapping criterion for the test items and procedure for estimating the cut scores that will result in a good representation of the intended standard on the reporting score scale. The requirements for the level of detail of PLDs also need investigation. Because there are many common components among the Bookmark, ESS, and ID Matching kernels, it is important to know how the standards estimated using these methods differ, and if they can be shown to accurately represent the intentions of the agency that called for the standard.

10.5 Standard Setting on New Test Designs

The initial standard setting kernels for collecting information from panelists about their estimates of performance of minimally qualified individuals on tests and items were used with tests composed of multiple-choice, or other dichotomously scored, test items. The Nedelsky (1954), Angoff (1971), and Bookmark (Lewis, Mitzel, & Green, 1996) kernels all were first applied to tests composed of multiple-choice items. Test designs changed in the 1990s to include more constructed-response and essay items that were human-scored using rating scales. Some of these tests required examinees to generate essay responses as well as extended responses to specific questions. As a result, the typically used procedures for setting standards were revised to be appropriate for these extended response items. Hambleton and Pitoniak (2006) described an extension of the Angoff probability-estimation kernel that uses estimated mean scores on rubrics rather than probabilities of correct responses on dichotomous items. Also, the Angoff item score kernel, the Yes/No procedure, was extended to use the number of points the minimally qualified examinee will get on the rubric rather than the mean value. Boursicot, Roberts, and Pell (2006) questioned whether these extensions of the Angoff kernels are appropriate for performance assessments. Hambleton (1998) had similar concerns and suggested an alternative method he called the "paper selection method." For this method, panelists were provided with a set of examples of student work that cover the range of the rating scale used for evaluating the work. They were asked to select papers that would be produced by the minimally qualified student. The selections were done for each performance assessment item in the test. The average score assigned to the selected papers was the estimate of the standard. If there are multiple open-ended items on the test, the sum of the average scores was the estimate of the standard. Hambleton (1998) indicated that the process was time consuming and tedious, and that it might result in small numbers of selected papers. Plake (1998) described some variations on the paper selection method. She suggested that panelists select two papers from each task that either represent the minimally acceptable work or bracket that point. The results were aggregated to get the standard on the full test. She indicated that the quality and distribution of the papers used in the process is critical.

Other alternatives have been proposed for setting standards on complex, open-ended responses. One is called the "body of work method." Pitoniak and Cizek (2016) described the method as one that involved sorting student work into categories that demonstrate above the minimal level of quality and below that level. As with all standard setting methods, there are variations in this method. Sometimes there are multiple rounds that use many examples of student work. In other implementations, there is a "range finding" round and then a "pinpointing" round. In the pinpointing round, papers are selected to be close to the expected cut score. Once the papers are classified, logistic regression is used to estimate the cut score. If there are multiple cut scores, ordinal logistic regression can be used if an assumption of equal slopes at all cut scores makes sense.

With the increased use of computer technology for testing, there has been an increase in the development of new assessment tasks that take advantage of the computing power. For example, Karakolides, O'Leary, and Scully (2021) investigated the use of animation to develop assessments that would reduce the impact of written text on the measurement of the target construct. A book edited by Jiao and Lissitz (2018) provided a cross section of some of the advances in item tasks and scoring procedures in development at that time. The book contains chapters that describe the uses of virtual reality, games, and simulations for assessment, as well as the challenges of scoring these assessments. The scoring of the assessments is a nontrivial problem because of the capability of the computer to record timing information and every keystroke and mouse click. As an example, Cui, Guo, Leighton, and Chu (2020) reported on research to analyze the full log file from a technology-based assessment to get information about student problem-solving procedures.

These new assessment types provide challenges to the standard setting processes because the test tasks are very complex, and the scoring process may yield results that do not have simple, intuitive meanings like "correct" and "incorrect." Computer scoring of essays and analyses of log files may result in scores on abstract continuous scales that require statistical sophistication for proper interpretation. It is not clear how current standard setting kernels can be adapted to obtain information from panelists for the estimation of a cut score on the reporting score scale for such assessments. It may be that procedures based on the reported test score for the assessments, such as the contrasting groups kernel, will need to be used to estimate standards on the scales resulting from these innovative test designs. There is a clear need for research and development on standard setting methods for use on these types of assessments.

10.6 Some Final Thoughts

Standards of performance have always been with us. Prior to the 1950s, performance standards were mostly in the heads of persons making decisions.

Selection for jobs or college admissions may have used tests with an arbitrary decision such as selecting the top 5% or some other percentage. With the standards-based education movement in the 1970s and legal requirements related to fairness in hiring, it became important to document how standards were set and whether the process was logical and replicable. Those pressures led to the development of formal methodology for setting standards and thorough documentation of the process that was used.

An important consideration as the process for setting performance standards became more formalized was who "owned" the standard? Who had the responsibility of providing information about the credibility of the standards and defending them against criticisms? There was some debate about who actually sets the standard. Was the committee who convened to work through the standard setting process the group who set the standard or was that committee only recommending the standard to a policy body who ultimately decides on the standard? It had been argued that standards are a matter of expert opinion, and the credibility of the standards rests upon the credentials of the experts involved in the process. I once attended a meeting where an educational policy maker said that he could set the standard, thus saving time and money by eliminating an extended process. He considered himself as an expert with excellent credentials, so his opinion was sufficient for setting the performance standard on an educational test. I thought of this facetiously as the "great person method" of standard setting. The field of standard setting has moved away from that perspective. Now it is generally accepted that there is a policy group, called the agency in this book, who is responsible for the standard and who ultimately sets the standard. The result of a standard setting process is considered as a recommendation to this policy group, and they can choose to deviate from the recommendation if they decide it does not match their intentions.

Accepting that the policy group (the agency) that called for the standard is the organization that makes the final decision about the location of the standard has led to the position taken in this book that the standard setting process is one of translation of policy rather than one of selecting experts to set a standard. Considering the standard setting process as one of translation of policy means that the agency can review the recommendations from the process and determine if the recommendations match what they intended by the policy. If the recommendations do not match the policy, it is legitimate for the agency to adjust the standard to improve the match to policy or require that the standard setting process be redone. If the standard setting process were considered as collecting the opinions of experts, the agency would have to argue that those opinions were incorrect in some way—a position that is much more difficult to justify.

Accepting that standard setting is a process of translating policy raises the issue of translating from what language to what other language. The language of policy is obvious. It is the written language that describes expectations for those who meet the standard. The "other language" is less obvious.

It is the language of the reporting score scale for the test. That reporting score scale is defined by the process of test construction—item writing and item selection—as well as the method for computing the reported score. Those who design and implement standard setting methodologies are beginning to recognize the importance of the test development process. Both the ESS and ID Matching methods described earlier in this chapter emphasize the characteristics of the test. Work on NAEP led to recognition that it was difficult to set standards at the upper and lower ends of the reporting score scale if there were no items assessing skills and knowledge at those levels. Standard setting practitioners now recommend that content definitions for standards be produced early in the process so that they can be used to guide the test design process. The meaning of the reported test score scale is critical. That score scale is the target language for the translation process. The reporting score scale is the operationalization of the continuum implied by the policy definition produced by the agency.

Standard setting is the translation of policy. There is a written language version of the policy and there is a target language for operational implementation of policy that is the language of the reporting score scale for an appropriately designed test. Under this framework, standard setting methods are a means of collecting information from qualified translators that will allow accurate estimation of a point on the reporting score scale that corresponds to the written policy. The task for those working in the area of standard setting is to determine the characteristics of the information to be collected from translators and how that information can be used to estimate the point on the reporting score scale. The many methods that have been suggested for standard setting highlight the creativity of researchers. The many methods and the modifications to the methods are both a blessing and a curse. They are a blessing because they allow the match of methods to the specific characteristics of the policy and the tests (e.g., dichotomous versus extended open-ended items, one standard versus several standards on the same reporting score scale). They are a curse in that different methods give different results and there is no clear guidance about how to select the method that will best represent the intentions of the agency.

One goal of this book has been to provide a framework for thinking about standard setting that will allow the evaluation of standard setting methods to determine which are better at recovering the intentions of the agency than others. Translation processes are challenging because any two languages do not represent ideas in exactly the same way. That means that equally qualified translators can produce slightly different translations that are equally good. It does not mean that all translations are equally good. Developers of and researchers into the standard setting processes must continue to evaluate the procedures that are being used to determine how well they recover the intentions implied by policy statements.

References

American Educational Research Association (AERA), American Psychological Association (APA) & National Council on Measurement in Education (NCME). (1999). *Standards for educational and psychological testing.* Washington, DC: American Educational Research Association.

American Educational Research Association (AERA), American Psychological Association (APA) & National Council on Measurement in Education (NCME). (2014). *Standards for educational and psychological testing.* Washington, DC: American Educational Research Association.

Angoff, W. H. (1971). Scales, norms, and equivalent scores. In R. L. Thorndike (Ed.), *Educational measurement* (pp. 508–600) (2nd ed.). Washington, DC: American Council on Education.

Berk, R. A. (1986). A consumer's guide to setting performance standards on criterion-referenced tests. *Review of Educational Research, 56,* 137–172.

Beuk, C. H. (1984). A method for reaching a compromise between absolute and relative standards in examinations. *Journal of Educational Measurement, 21*(2), 147–152.

Birnbaum, A. (1968). Some latent trait models and their use in inferring an examinee's ability. In F. M. Lord & M. R. Novick (Eds.), *Statistical theories of mental test scores.* (pp. 397–479) Reading, MA: Addison-Wesley.

Bourque, M. L. (2009, March). *A history of NAEP achievement levels: Issues, implementation, and impact 1989–2009.* Paper commissioned for the 20th anniversary of the National Assessment Governing Board 1988–2008.

Bourque, M. L., & Webb, L. (2011). *Finalize national assessment of educational progress (NAEP) writing achievement levels descriptions (ALDs).* Leesburg, VA: Mid-Atlantic Psychometric Services, Inc.

Boursicot, K. A. M., Roberts, T. E., & Pell, G. (2006). Standard setting for clinical competence at graduation from medical school: A comparison of passing scores across five medical schools. *Advances in Health Sciences Education, 11,* 173–183.

Brennan, R. L. (2002). *Estimating standard error of a mean when there are only two observations.* (CASMA Technical Note No. 1). Iowa City, IA: University of Iowa Center for Advanced Studies in Measurement and Assessments.

Brennan, R. L., & Kane, M. T. (1977). An index of dependability for mastery tests. *Journal of Educational Measurement, 14*(3), 277–289.

Brennan, R. L., & Lockwood, R. E. (1980). A comparison of the Nedelsky and Angoff cutting score procedures using generalizability theory. *Applied Psychological Measurement, 4*(2), 219–240.

Burt, W. M., & Stapleton, L. M. (2010). Connotative meanings of student performance labels used in standard setting. *Educational Measurement: Issues and Practice, 29*(4), 28–38.

Cavell, S. A. (2010). A social history of midshipmen and quarterdeck boys in the Royal Navy, 1761–1831. Unpublished doctoral dissertation. Exeter, England: University of Exeter.

Cizek, G. J. (1996). Standard-setting guidelines. *Educational Measurement: Issues and Practice, 15*(1), 13–21.

Cizek, G. J. (2012). *Setting performance standards: Foundations, methods, and innovations* (2nd ed.). New York: Routledge.

Cizek, G. J. (2012a). An introduction to contemporary standard setting. In G. J. Cizek (Ed.), *Setting performance standards: Foundations, methods, and innovations* (pp. 3–14) (2nd ed.). New York: Routledge.

Cizek, G. J. (2012b). The forms and functions of evaluations in the standard setting process. In G. J. Cizek (Ed.), *Setting performance standards: Foundations, methods, and innovations* (pp. 165–178) (2nd ed.). New York: Routledge.

Cizek, G. J., & Bunch, M. B. (2007). *Standard setting: A guide to establishing and evaluating performance standards on tests.* Thousand Oaks, CA: Sage.

Clauser, J. C., Clauser, B. E., & Hambleton, R. K. (2014). Increasing the validity of Angoff standards through analysis of judge-level internal consistency. *Applied Measurement in Education, 27*(1), 19–30.

Clauser, B. E., Kane, M., & Clauser, J. C. (2020). Examining the precision of cut scores within a generalizability theory framework: A closer look at the item effect. *Journal of Educational Measurement, 57*(2), 216–229.

Clauser, J. C., Margolis, M. J., & Clauser, B. E. (2014). An examination of the replicability of Angoff standard setting results within a generalizability theory framework. *Journal of Educational Measurement, 51*(2), 127–140.

Clauser, J. C. Mee, J., & Margolis, M. J. (2013). The effect of data format on integration of performance data into Angoff judgments. *International Journal of Testing, 13*(1), 65–85.

Clauser, B. E., Swanson, D. B., & Harik, P. (2002). Multivariate generalizability analysis of the impact of training and examinee performance information on judgments made in an Angoff-style standard-setting procedure. *Journal of Educational Measurement, 39*(4), 269–290.

Cohen, A. S., Kane, M. T., & Crooks, T. J. (1999). A generalized examinee-centered method for setting standards on achievement tests. *Applied Measurement in Education, 12*(4), 343–366.

Cohen, Y. (2017). Estimating the intra-rater reliability of essay raters. *Frontiers in Education*, 1–11. https://doi.org/10.3389/feduc.2017.00049.

Conaway, L. E. (1979). Setting standards in competency-based education: Some current practices and concerns. In M. A. Bunda & J. R. Sanders (Eds.), *Practices and problems in competency-based measurement.* (pp. 72–88) Washington, DC: National Council on Measurement in Education.

Council of Europe. (2001). *A common European framework of reference for learning, teaching and assessment.* Cambridge, England: Cambridge University Press.

Council of Europe. (2009). *Relating language examinations to the common European framework of reference for languages: Learning, teaching, assessment (CEFR). A manual.* Strasbourg, France: Council of Europe Language Policy Division.

Cui, Y., Guo, Q., Leighton, J. P., & Chu, M-W. (2020). Log data analysis with ANFIS: A fuzzy neural network approach. *International Journal of Testing, 20*(1), 78–96.

Cusimano, M. D. (1996). Standard setting in medical education. *Academic Medicine, 71*(10) Supplement, S112–S120.

Dorsey, C. F., & Schowalter, J. M. (2008). *The first 25 years: 1978–2003.* Chicago, IL: National Council of State Boards of Nursing.

Edison Electric Institute. (2001). *Power plant maintenance positions selection system: Test brochure*. Washington, DC: Author.

Egan, K. L., Schneider, M. C., & Ferrara, S. (2012). Performance level descriptors: History, practice, and a proposed framework. In G. J. Cizek (Ed.), *Setting performance standards: Foundations, methods, and innovations* (Pp. 79–106) (2nd ed.). New York: Routledge.

Everson, H. T., & Forte, E. (2020). Introduction to special section: Issues and advances in standard setting methods. *Educational Measurement: Issues and Practices, 39*(1), 7.

Ferrara, S., Johnson, E., & Chen, W.-H. (2005). Vertically articulated performance standards: Logic, procedures, and likely classification accuracy. *Applied Measurement in Education, 18*(1), 35–60.

Ferrara, S., Perie, M., & Johnson, E. (2002). *Matching the judgmental task with standard setting panelist expertise: The item-descriptor (ID) matching method*. Paper presented at the annual meeting of the National Council on Measurement in Education, New Orleans, LA.

Ferrara, S., Svetina, D., Skucha, S., & Davidson, A. H. (2011). Test development with performance standards and achievement growth in mind. *Educational Measurement: Issues and Practice, 30*(4), 3–15.

Flexner, S. B. (Ed.). (1987). *The random house dictionary of the English language* (2nd ed.). New York: Random House.

Florida Department of Education. (2020). *Florida statewide science and EOC assessments 2020 technical report*. Tallahassee, FL: Author.

Glass, G. V. (1978). Standards and criteria. *Journal of Educational Measurement, 15*(4), 237–261.

Guttman, L. (1950). The basis for scalogram analysis. In S. A. Stouffer et al. (Eds.), *Measurement and prediction. Studies in social psychology in World War II, Vol. 4* (pp. 60–90). Princeton, NJ: Princeton University Press.

Haertel, E. (1985). Construct validity and criterion referenced testing. *Review of Educational Research, 55*(1), 23–46.

Hambleton, R. K. (1998). Setting performance standards on achievement tests: Meeting the requirements of title I. In L. N. Hansche (Ed.), *Handbook for the development of performance standards* (pp. 87–114) Washington, DC: Council of Chief State School Officers.

Hambleton, R. K., & Pitoniak, M. J. (2006). Setting performance standards. In R. L. Brennan (Ed.), *Educational measurement* (pp. 433–470) Westport, CT: American Council on Education and Praeger.

Hansche, L. N. (1998). *Handbook for the development of performance standards: Meeting the requirements of title I*. Washington, DC: Council of Chief State School Officers and Office of Elementary and Secondary Education.

Harrison, G. M. (2015). Non-numeric intrajudge consistency feedback in an Angoff procedure. *Journal of Educational Measurement, 52*(4), 399–418.

Hartka, E. (1993). Comparisons of student performance on NAEP and other standardized tests. In Shepard, L. Glaser, R., Linn, R., & Bohrnstedt, R. (Eds.), *Setting performance standards for student achievement: A report of the National Academy of Education on the evaluation of the NAEP trial state assessment: An evaluation of the 1992 achievement levels* (pp. 175–177) Stanford, CA: The National Academy of Education, Stanford University, School of Education.

Hofstee, W. K. B. (1983). The case for compromise in educational selection and grading. In S. B. Anderson & J. S. Helmick (Eds.), *On Educational Testing* (pp. 109–127) San Francisco, CA: Jossey-Bass.

Hollenbeck, K., Tindal, G., & Almond, P. (1999). Reliability and decision consistency: An analysis of writing mode at two times on a statewide test. *Educational Assessment*, *6*(1), 23–40.

Hoyt, C. (1941). Test reliability estimated by analysis of variance. *Psychometrika*, *6*(3), 153–160.

Hsieh, M. (2013). Comparing yes/no Angoff and bookmark standard setting methods in the context of English assessment. *Language Assessment Quarterly*, *10*(3), 331–350,

Hunter, J. (2011). *Preparing students for the world beyond the classroom: Linking EQAO assessments to 21st-century skills* (Research Bulleting 7). Toronto, ON: Education Quality and Accountability Office.

Jaeger, R. M. (1989). Certification of student competence. In R. L. Linn (Ed.), *Educational measurement* (pp. 485–514) (3rd ed.). New York: American Council on Education and Macmillan.

Jaeger, R. M. (1990). Establishing standards for teacher certification tests. *Educational Measurement: Issues and Practices*, *9*(4), 15–20.

Jaeger, R. M. (1995). Setting performance standards through two-stage judgmental policy capture. *Applied Measurement in Education*, *8*(1), 15–40.

Jiao, H., & Lissitz, R. W. (Eds.). (2018). *Technology enhanced innovative assessment: Development, modeling, and scoring from an interdisciplinary perspective*. Charlotte, NC: Information Age Publishing.

Kahl, S. R., Crockett, T. L., DePascale, C. A., & Rindfleisch, S. L. (1994, June). *Using actual student work to determine cutscores for proficiency levels: New methods for new tests*. Paper presented at the National Conference on Large-Scale Assessments, Albuquerque, NM.

Kane, M. (1994). Validating the performance standards associated with passing scores. *Review of Educational Research*, *64*(3), 425–461.

Kane, M. (2001). So much remains the same: Conceptions and status of validation in setting standards. In G. Cizek (Ed.), *Standard setting: Concepts, methods, and perspectives* (pp. 53–88). Mahwah, NJ: Erlbaum.

Kane, M. (2013). Validating the interpretations and uses of test scores. *Journal of Educational Measurement*, *50*(1), 1–73.

Karakolidis, A., O'Leary, M., & Scully, D. (2021). Animated videos in assessment: Comparing validity evidence from the test-takers' reactions to an animated and a text-based situational judgment test. *International Journal of Testing*, *21*(2), 57–79.

Kingston, N. M., & Tiemann, G. C. (2012). Setting performance standards on complex assessments: The body of work method. In G. J. Cizek (Ed.), *Setting performance standards: Foundations, methods, and innovations* (pp. 201–224) (2nd ed.). New York: Routledge.

Lehmann, E. L., & Casella, G. (1998). *Theory of point estimation* (2nd ed.). New York: Springer.

Leucht, R. M. (2006). Designing tests for pass-fail decisions using item response theory. In S. M. Downing & T. M. Haladyna (Eds.), *Handbook of test development*. (pp. 575–596) Mahwah, NJ: Erlbaum.

Lewis, D. M. (2002). *Standard setting with vertical scales*. Paper presented at the annual meeting of the National Council on Measurement in Education, New Orleans, LA.

Lewis, D. M., & Cook, R. (2020). Embedded standard setting: Aligning standard-setting methodology with contemporary assessment design principles. *Educational Measurement: Issues and Practice, 39*(1), 8–21.

Lewis, D. M., Green, D. R., Mitzel, H. C., Baum, K., & Patz, R. J. (1998). *The bookmark standard setting procedure: Methodology and recent implementations.* Paper presented at the 1998 annual meeting of the National Council on Measurement in Education, San Diego, CA.

Lewis, D. M., & Haug, C. A. (2005). Aligning policy and methodology to achieve consistent across-grade performance standards. *Applied Measurement in Education, 18*(1), 11–34.

Lewis, D. M., Mitzel, H. C., & Green, D. R. (1996). *Standard setting: A bookmark approach "cutpoints through bookmarks".* Paper presented at the CCSSO National Conference on Large Scale Assessment, Phoenix, AZ.

Linn, R. L., Baker, E. L., & Betebenner, D. W. (2002). Accountability systems: Implications of requirements of the No Child Left Behind Act of 2001. *Educational Researcher, 31*(6), 3–16.

Linn, R. L., Madaus, G., & Pedulla, J. (1982). Minimum competency testing: Cautions on the state of the art. *American Journal of Education, 91*, 1–35.

Lissitz, R. W., & Huynh, H. (2002). Vertical equating for state assessments: Issues and solutions in determination of adequate yearly progress and school accountability. *Practical Assessment, Research, and Evaluation, 8*(10), 1–6.

Livingston, S. A. (1972). A criterion-referenced application of classical test theory. *Journal of Educational Measurement, 9*(1), 13–26.

Livingston, S., & Zieky, M. (1982). *Passing scores: A manual for setting standards of performance on educational and occupational tests.* Princeton, NJ: Educational Testing Service.

Loomis, S. C., & Bourque, M. L. (2001). From tradition to innovation: Standard setting on the National Assessment of Educational Progress. In G. J. Cizek (Ed.), *Setting performance standards: Concepts, methods, and perspectives.* (pp. 175–217) Mahwah, NJ: Lawrence Erlbaum.

Loomis, S. C., Hanick, P. L., Bay, L., & Crouse, J. D. (2000). *Developing achievement levels for the 1998 NAEP in writing interim report: Field trials.* Iowa City, IA: ACT Inc.

Loomis, S. C., Hanick, P. L., & Yang, W.-L. (2000). *Setting achievement levels on the 1998 national assessment of educational progress in writing: ALS final report.* Iowa City, IA: ACT Inc.

Lord, F. M. (1980). *Applications of item response theory to practical testing problems.* Hillside, NJ: Lawrence Erlbaum.

Margolis, M. J., Mee, J., Clauser, B. E., Winward, M., & Clauser, J. C. (2016). Effect of content knowledge on Angoff-style standard setting judgments. *Educational Measurement: Issues and Practice, 35*(1), 29–37.

McLaughlin, D. H. (1993). Order of Angoff ratings in setting multiple simultaneous standards. In Shepard, L. Glaser, R., Linn, R., & Bohrnstedt, R. (Eds.), *Setting performance standards for student achievement: A report of the national academy of education on the evaluation of the NAEP trial state assessment: An evaluation of the 1992 achievement levels.* (pp. 158–161) Stanford, CA: The National Academy of Education, Stanford University, School of Education.

Michigan Department of Education. (2010). *Michigan K-12 standards: English language arts.* Lansing, MI: Author.

Mills, C. N., & Jaeger, R. M. (1998). Creating descriptions of desired student achievement when setting performance standards. In N. L. Hansche (Ed.), *Handbook for the development of performance standards: Meeting the requirements of title I.* (pp. 73–85) Washington, DC: Council of Chief State School Officers and Office of Elementary and Secondary Education.

Mitzel, H. C., Lewis, D. M., Patz, R. J., & Green, D. R. (2001). The bookmark procedure: Psychological perspectives. In G. J. Cizek (Ed.), *Setting performance standards: Concepts, methods, and perspectives* (pp. 249–281). Mahwah, NJ: Lawrence Erlbaum Associates.

National Assessment Governing Board. (2018). *Developing student achievement levels for the National Assessment of Educational Progress: Policy statement.* Washington, DC: Author.

National Assessment Governing Board. (2019). *Mathematics framework for the 2019 National Assessment of Educational Progress.* Washington, DC: Author.

National Center for Education Statistics. (2014). *A first look: 2013 Mathematics and reading – National Assessment of Educational Progress at grades 4 and 8 (NCES 2014-451).* Washington, DC: Institute for Education Sciences, U.S. Department of Education.

National Council of State Boards of Nursing. (2011). *What you need to know about nursing licensure and boards of nursing.* Chicago, IL: Author.

Nedelsky, L. (1954). Absolute grading standards for objective tests. *Educational and Psychological Measurement, 14*(1), 3–19.

Nunnally, J. C., & Bernstein, I. H. (1994). *Psychometric theory* (3rd ed.). New York: McGraw-Hill.

Peabody, M. R., & Wind, S. A. (2019). Exploring the influence of judge proficiency on standard-setting judgments. *Journal of Educational Measurement, 56*(1), 101–120.

Pearson, D., & DeStafano, L. (1993). An evaluation of the 1992 NAEP reading achievement levels, report three: Comparison of cutpoints for the 1992 NAEP reading achievement levels with those set by alternate means. In Shepard, L. Glaser, R. Linn, R., & Bohrnstedt, R. (Eds.), *Setting performance standards for student achievement: A report of the National Academy of Education Panel on the evaluation of the NAEP trial state assessment: An evaluation of the 1992 achievement levels.* Stanford, CA: The National Academy of Education, Stanford University, School of Education.

Pellegrino, J. W., Jones, L. R., & Mitchell, K. J. (Eds.). (1999). *Grading the nations' report card: Evaluating NAEP and transforming the assessment of educational progress.* Washington, DC: National Academy Press.

Perie, M. (2008). A guide to understanding and developing performance-level descriptors. *Educational Measurement: Issues and Practice, 27*(4), 15–29.

Phillips, S. E. (2010). *Assessment law in education.* Phoenix: S.E. Phillips.

Pitoniak, M. J., & Cizek, G. J. (2016). Standard setting. In C. S. Wells & M. Faulkner-Bond (Eds.), *Educational measurement: From foundations to future* (pp. 38–61). New York: Guilford Press.

Plake, B. S. (1998). Setting performance standards for professional licensure and certification. *Applied Measurement in Education, 11*(1), 65–80.

Plake, B. S., & Cizek, G. J. (2012). Variations on a theme: The modified Angoff, extended Angoff, and yes/no standard setting methods. In G. J. Cizek (Ed.), *Setting performance standards: Foundations, methods, and innovations* (2nd ed.) (pp. 181–199). New York: Routledge.

Plake, B. S., Melican, G. J., & Mills, C. N. (1991). Factors influencing intrajudge consistency during standard-setting. *Educational Measurement: Issues and Practice, 10*(1), 15–16, 22, 25–26.

Rasch, G. (1960). *Probabilistic model for some intelligence and achievement tests.* Copenhagen: Danish Institute for Educational Research.

Raymond, M. R., & Reid, J. B. (2001). Who made thee a judge? Selecting and training participants for standard setting. In G. Cizek (Ed.), *Standard setting: Concepts, methods, and perspectives* (pp. 119–158). Mahwah, NJ: Erlbaum.

Reckase, M. D. (2000). *Evolution of the NAEP achievement levels setting process: A summary of the research and development efforts conducted by ACT.* Iowa City, IA: ACT, Inc.

Reckase, M. D. (2006). A conceptual framework for a psychometric theory for standard setting with examples of its use for evaluating the functioning of two standard setting methods. *Educational Measurement: Issues and Practice, 25*(2), 4–18.

Reckase, M. D. (2017). *A tale of two models: Sources of confusion in achievement testing* (Research Report No. RR-17-44). Princeton, NJ: Educational Testing Service.

Schneider, M. C., Huff, K. L., Egan, K. L., Gaines, M. L., & Ferrara, S. (2013). Relationships among item cognitive complexity, contextual demands, and item difficulty: Implications for achievement-level descriptors. *Educational Assessment, 18*(2), 99–121.

Schulz, E. M., & Mitzel, H. C. (2006, April). *The mapmark standard-setting method.* Paper presented at the annual meeting of the National Council of Measurement in Education, Montreal, QC.

Scriven, M. (1978). How to anchor standards. *Journal of Educational Measurement, 15*(4), 273–275.

Shepard, L. (1980). Standard setting issues and methods. *Applied Psychological Measurement, 4*(4), 447–467.

Shepard, L., Glaser, R., Linn, R., & Bohrnstedt, G. (1993). *Setting performance standards for student achievement: A report of the National Academy of Education on the evaluation of the NAEP trial state assessment: An evaluation of the 1992 achievement levels.* Stanford, CA: The National Academy of Education, Stanford University School of Education.

Sireci, S. G., Hambleton, R. K., & Pitoniak, M. J. (2004). Setting passing scores on a licensure examinations using direct consensus. *CLEAR Exam Review, 15*(1), 21–25.

Skaggs, G., Hein, S. F., & Wilkins, J. L. M. (2020). Using diagnostic profiles to describe borderline performance in standard setting. *Educational Measurement: Issues and Practices, 39*(1), 45–51.

Stout, W. F. (1987). A nonparametric approach for assessing latent trait dimensionality. *Psychometrika, 52*(4), 589–617.

Surowiecki, J. (2004). *The wisdom of crowds.* New York: Anchor Books.

Teng, S.-Y. (1943). Chinese influence on the Western examination system: I. Introduction. *Harvard Journal of Asiatic Studies, 7*(4), 267–312.

The Mathworks, Inc. (2022). *MATLAB: The language of technical computing.* Natick, MA: Author.

Thorndike, E. L., Bregman, E. O., & Cobb, M. V. (1924). The selection of tasks of equal difficulty by a consensus of opinion. *Journal of Educational Research, 9*, 133–139.

Tinkelman, S. (1947). *Difficulty prediction of test items.* New York: Bureau of Publications, Teachers College, Columbia University.

Truxillo, D. M., Donahue, L. M., & Sulzer, J. L. (1996). Setting cutoff scores for personnel selection tests: Issues, illustrations, and recommendation. *Human Performance, 9*(3), 275–295.

van der Linden, W. J. (1982). A latent trait method for determining intrajudge inconsistency in the Angoff and Nedelsky techniques of standard setting. *Journal of Educational Measurement, 19*(3), 295–308.

van der Linden, W. J. (1995). A conceptual analysis of standard setting in large-scale assessments. In *Proceedings of Joint Conference on Standard Setting for Large-Scale Assessment of the National Assessment Governing Board (NAGB) and the National Center for Education Statistics (NCES).* Washington, DC: National Assessment Governing Board.

Vinovskis, M. A. (1998). *Overseeing the nation's report card: The creation and evolution of the National Assessment Governing Board (NAGB).* Washington, DC: National Assessment Governing Board, U. S. Department of Education.

Wainer, H., & Mislevy, R. J. (2000). Item response theory, item calibration, and proficiency estimation. In H. Wainer (Ed.), *Computerized adaptive testing: A primer* (pp. 61–100) (2nd ed.). Mahwah, NJ: Lawrence Erlbaum Associates.

Wang, N. (2003). Use of the Rasch IRT model in standard setting: An item mapping method. *Journal of Educational Measurement, 40*(3), 231–253.

Webb, N. L. (2006). Identifying content for student achievement tests. In S. M. Downing & T. M. Haladyna (Eds.), *Handbook of test development.* Mahwah, NJ: Lawrence Erlbaum Associates.

Wolkowitz, A. A., Impara, J. C., & Buckendahl, C. W. (2017). Closing the loop: Providing test developers with performance level descriptors so standard setters can do their job. In S. Blömeke & J. E. Gustafsson (Eds.), *Standard setting in education. Methodology of educational measurement and assessment.* Cham, Switzerland: Springer.

Wyse, A. E. (2018). Regression effects in Angoff ratings: Examples from credentialing exams. *Applied Measurement in Education, 31*(1), 68–78.

Wyse, A. E., & Reckase, M. D. (2012). Examining rounding rules in Angoff-type standard-setting methods. *Educational and Psychological Measurement, 72*(2), 224–244.

Yang, W.-L. (2000, April). *Analysis of item ratings for ensuring the procedural validity of the 1998 NAEP achievement-levels setting.* Paper presented as the annual meeting of the American Educational Research Association, New Orleans, LA.

Index

R

S

T

Printed in the United States
by Baker & Taylor Publisher Services